Metal Ions in Biology

This book discusses the inherent need for and significance of metal ions in metabolic reactions. It details their essential elements and mechanistic prospects in regulating biological reactions, as well as covers the broad domain of elements, including main group and transition elements, in a comparative and comprehensive manner. Contemporary and advanced topics, such as nuclear medicine and biomineralization, are also covered.

Features:

- Highlights the different behaviors of Fe/Cu proteins and structural changes during their biological functioning.
- Explains the mechanistic prospects of various enzymes and proteins, e.g., Cu-Zn SOD, zinc finger, and ionophores.
- Explores the chemical and biological prospects of trace and ultra-trace elements.
- Includes biomimetic models of iron and copper.
- Reviews the criteria for nature selecting metal ions, why only certain elements are essential, and the differences between biologically induced and biologically controlled biomineralization.

This book is aimed at graduate students and researchers in chemical engineering, materials science, chemistry, and biological sciences.

Metal Ions in Biology

Adish Tyagi, Rohit Singh Chauhan, and
A. K. Tyagi

CRC Press
Taylor & Francis Group
Boca Raton London New York

CRC Press is an imprint of the
Taylor & Francis Group, an **informa** business

First edition published 2025
by CRC Press
2385 NW Executive Center Drive, Suite 320, Boca Raton FL 33431

and by CRC Press
4 Park Square, Milton Park, Abingdon, Oxon, OX14 4RN

CRC Press is an imprint of Taylor & Francis Group, LLC

ISBN: 9781032458823 (hbk)
ISBN: 9781032558097 (pbk)
ISBN: 9781003432333 (ebk)

DOI: 10.1201/9781003432333

Typeset in Times
by codeMantra

Contents

Preface

Bioinorganic chemistry is a leading science discipline at the interface of chemistry and biology. This book, *Metal Ions in Biology*, delves into the captivating interplay between metals and living systems, unraveling the chemical underpinnings of essential biological processes. Metal ions play pivotal roles, contributing to a multitude of biological functions such as respiration, electron transport, oxygen and nitrogen transport, photosynthesis, nerve impulses, muscle contractions, and defense mechanisms against toxins and mutagenic agents. The importance of metal ions in maintaining the life cycles of animals and plants cannot be overstated. Moreover, metal ions find extensive applications in the realms of diagnosis and treatment, a testament to the profound interface between chemistry and biology within living organisms.

Beginning with an introduction to metal ions, this book highlights the interdisciplinary nature of the subject and sheds light on the role of metal ions in biological systems. It explores why certain metal ions are selectively used in living organisms, how nature makes these choices, and the biological ligands employed to introduce metal ions into biological systems. Furthermore, it delves into the harmful consequences of both excess and deficiency of metal ions on human health. A chapter is dedicated to the role of coordination chemistry and its applications in biological systems. After laying these foundational concepts, this book embarks on a journey to understand and elucidate the pivotal functions of important metal ions, which constitute a significant portion of university curricula across India. As this book approaches its conclusion, it discusses the crucial role of metal ions in medicine and radiopharmaceuticals. Additionally, there is a dedicated chapter on biomineralization. Throughout, this book strives to present these fascinating topics in a lucid and straightforward manner.

The authors extend their sincere gratitude to their colleagues and students, with special recognition for Saili Lokhande, Atharva Kulkarni, Suraj Yadav, Shraddha Yadav, and Aditya Bhatt. These individuals dedicated their time to proofreading the manuscript and providing valuable suggestions for its enhancement. Furthermore, the authors express their appreciation and gratitude to their publisher for their unwavering support and patience in ensuring the quality of production.

The authors truly believe that the present edition of this book will be well received by students and teachers alike. All suggestions for enhancing the quality of the text are most welcome.

Author Biographies

Adish Tyagi works as Scientific Officer-F in the Chemistry Division of the Bhabha Atomic Research Centre (BARC), India. He pursued his postgraduation in chemistry from Delhi University, India, and obtained his Ph.D. in chemistry from Homi Bhabha National Institute, India. His research interest includes designing molecular precursors and their conversion into intricate functional nanostructures for energy conversion and storage applications. In addition to this, he is also working in the ultra-purification of materials. Towards teaching, Dr. Tyagi is a faculty at HBNI-Mumbai and visiting faculty at UM-DAE CEBS, Mumbai. He serves as an executive council member of the SMC and ICS(MB). He has published 56 research articles in peer-reviewed journals, five book chapters, six newsletters and guest edited four SMC-bulletins. He is the recipient of DAE-YASTA-2020, CRS Bronze Medal (2024), SMC Bronze Medal (2024), Young associate of Maharashtra Academy of Sciences, NASI Member, INYAS Member.

Rohit Singh Chauhan works as Dean, Research and Consultancy at K. J. Somaiya College of Science and Commerce, India. He pursued his post-graduation in Chemistry at Christ Church P. G. College, India, and was awarded Ph.D. from Bhabha Atomic Research Centre under the aegis of Homi Bhabha National Institute Deemed University, India. He has 50 international peer-reviewed papers to his credit and has been also awarded the Dr. S. K. Somaiya Award for Excellence in Research-Higher and Technical Education (2019), SMC Bronze Medal (2022), INSA-VSP award (2024), and recognition as Young Associate of Maharashtra Academy of Sciences. His research interests include synthetic inorganic chemistry, organometallics, and catalysis.

A. K. Tyagi joined Chemistry Division, BARC, Mumbai in 1986 through BARC Training School. During his career spanning close to four decades, he occupied various positions at BARC such as Director, Chemistry Group and Director, Bio-Science Group. He was a Distinguished Scientist, DAE. Presently, he is Dean and Senior Professor at Homi Bhabha National Institute (HBNI), Mumbai. He has to credit his 47 PhD students, about 680 papers in journals and 15 books. He has been conferred with a number of awards such as DAE-Homi Bhabha Science and Technology Award, DAE-SRC Outstanding Researcher Award, DAE-Group Achievement Award (four times); MRSI Medal; MRSI-ICSC Materials Science Senior Award; MRSI-CNR Rao Prize in Advanced Materials; MRSI Distinguished Materials Scientist of the year award; CRSI Bronze Medal; CRSI-CNR Rao National Prize in Chemical Sciences; CRSI-Silver Medal; Rajib Goyal Prize in Chemical Sciences; Metallurgist of the Year award from Ministry of Steel, GoI; National Prize in Solid State and Materials Chemistry from JNCASR; ISCA Acharya PC Ray Memorial Award; NETZSCH – ITAS Award by Indian Thermal Analysis Society, DN Agarwal Memorial Award from Indian Ceramics Society and Gold medal of Chirantan Rasayan Sanstha. In 2024, he was conferred with the prestigious Vigyan Shri Award of Government of India. He is a Fellow of several national and international Science and Engineering Academies such as National Academy of Sciences, India; Indian Academy of Sciences; Indian National Academy of Engineering; Indian National Science Academy; Royal Society of Chemistry, Asia Pacific Academy of Materials, World Academy of Ceramics, African Academy of Sciences and The World Academy of Sciences.

Introduction

Within the pages of *Metal Ions in Biology*, readers embark on an enthralling journey through the captivating realm of metal ions and their indispensable role in the intricacies of biological systems. It begins with an introductory chapter that illuminates the profound significance of metal ions in the very existence of life. It illuminates how nature has selected particular metal ions and incorporated them into the biological system. Additionally, this chapter highlights the detrimental effects associated with imbalances in metal ion concentrations, underscoring the delicate equilibrium required for the well-being of organisms. Further, this book ventures into the area of coordination chemistry, providing essential insights into understanding how metal ions perform their vital functions within biological systems. Dedicated chapters describe the function of specific metal ions and their importance for the existence of life from a chemistry perspective. As we progress through the chapters, this book shifts its focus from fundamental understanding to practical applications. It sheds light on the role of metal ions in diagnostics and therapeutics, where science meets medicine in a quest to enhance human health. The journey culminates with a revelation of nature's ingenuity—biomineralization and the inspiration that can be drawn from it, which could lead to groundbreaking advancements in diverse areas ranging from medicine and engineering to energy and environmental sustainability. *Metal Ions in Biology* is more than just a book; it is an odyssey through the heart of life's chemistry. It invites readers to explore the world of metal ions, where the seemingly ordinary becomes extraordinary, and the invisible threads of chemistry weave the tapestry of life itself.

1 Role of Metal Ions in Biology
An Overview

INTRODUCTION

Photosynthesis and respiration processes are considered the basis of life, and the role of metal ions in these processes is indispensable. Hence, it would be accurate to state that metal ions form the basis of life. Metal ions, complexes, and materials derived from them are essential components in all living systems. The study of the role of metal ions in biological systems falls under the scope of bioinorganic chemistry, serving as a bridge between inorganic chemistry and biochemistry. It explores the interdependence between the two sub-disciplines and mainly deals with the function of inorganic 'substances' in living systems. This includes the transport, speciation, and mineralization of inorganic materials. Furthermore, it encompasses the use of metal ions and their compounds in medicinal therapy and diagnosis. In the words of the Chairman of the '7th International Conference on Bioinorganic Chemistry' held in Lubeck, Germany, bioinorganic chemistry can be defined as the

> Chemistry devoted to all aspects of "inorganic elements" (such as transition metals) as being vital for the growth and metabolism of living systems. Bioinorganic Chemistry is a multidisciplinary field which draws on expertise in biochemistry, chemistry, crystallography, genetics, medicine, microbiology together with the effective application of advanced physical methods.

Chairman,
7th International Conference on Bioinorganic chemistry (1994)
Lubeck, GERMANY

From the very early days of human understanding of biological systems, organic chemistry was thought to be the primary chemistry operating in living systems, as biomolecules such as carbohydrates, fats, and proteins present in living systems are created by carbon-hydrogen bonds. However, it was gradually realized that metal ions play a vital role in a vast number of widely different life processes and they are not mere spectators in the organic world. Thus, it can be misleading to associate organic chemistry with life and inorganic chemistry solely with the inanimate world. In reality, life is built on organic and inorganic compounds. In this book, the significance of metal ions for life will be discussed, as these are usually excluded from the scope of organic chemistry.

DOI: 10.1201/9781003432333-1

Metal ions in biology, environment, and medicine play diverse and pivotal roles. From the movement of electrons in electron transfer pathways (photosynthesis and the respiratory chain of mitochondria), splitting water to produce oxygen, maintaining osmotic pressure, and transferring of electrical signals across the body in no time, metal ions constitute the active site of these important processes. Beyond this, metal ions in the form of complexes and materials are extensively used in life-saving drugs—example, platinum in cisplatin for anticancer therapy, gold compounds to reduce inflammation and slow disease progression in patients with rheumatoid arthritis, and lithium carbonate in the treatment of bipolar disorder. Paramagnetic metal complexes such as gadolinium complexes are widely used as contrast agents for magnetic resonance imaging (MRI) [1]. Moreover, it is clearly understood that a deficiency of certain essential metal ions can disrupt the proper functioning of the body, while excessive levels of metal ions can cause toxicity.

Referring to bioinorganic chemistry as a sub-discipline of chemistry is rather incorrect. It is, in fact, a truly multidisciplinary subject involving many branches of the physical, chemical, material, and biological sciences, among others, as depicted in Figure 1.1.

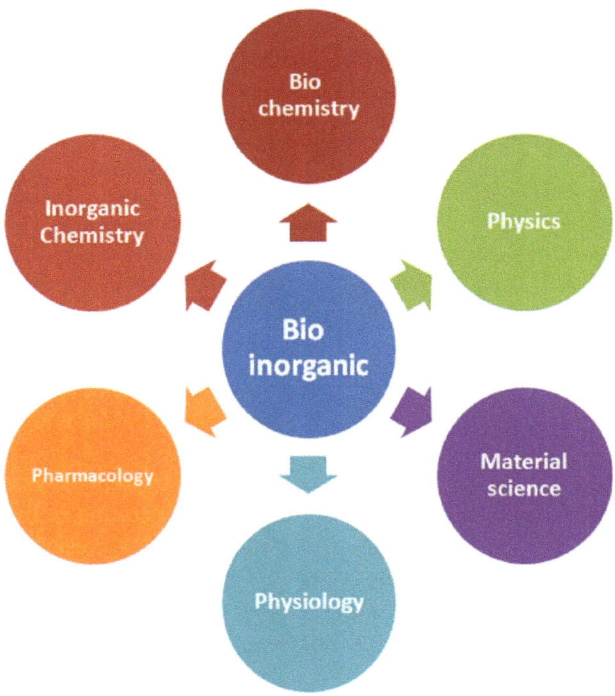

FIGURE 1.1 Multi-disciplinary nature of bioinorganic chemistry.

ESSENTIAL ELEMENTS FOR HUMAN SYSTEM

For an element to be classified as essential, it must meet some remarkable criteria. Firstly, it must be present in human tissue. Secondly, its absence from our system must produce irreversible detrimental effects on the body's vital functions. Furthermore, the reduction in the physiological function of the element can be normalized by supplementation of that element [2]. On this basis, around 25 elements out of the 118 elements in the periodic table are considered essential. In a 70 kg human body, the distribution of elements is given in Table 1.1.

The human body consists of 11 elements, with hydrogen, carbon, oxygen, nitrogen, calcium, and phosphorus accounting for 98.5% of its composition. The remaining 1.5% comprises sodium, potassium, magnesium, sulfur, and chlorine. While these six elements alone can construct essential biomolecules such as carbohydrates, fats, polypeptides, nucleic acids, and phospholipid bilayers, the question arises: why do we need metal ions at all? The answer lies in the fact that these organic building blocks alone are insufficient to support the body's functions. In addition to essential organic moieties, the human body requires enzymes and proteins to catalyze biochemical reactions. Metal compounds are indispensable components in many enzymes, playing vital roles that organic moieties cannot effectively fulfill. For example, redox reactions in the body rely on metal ions like iron and copper due to their superior electron transfer capabilities compared

TABLE 1.1
Amount of Element Present in Human Body Weighing 70 kg

Element	Weight (kg)
Oxygen	~45
Carbon	~13
Hydrogen	~7
Nitrogen	~2
Calcium	~1
Phosphorus	~0.7
Potassium	~0.1
Sodium	~0.1
Magnesium	~0.4
Iron	~0.04
Zinc	~0.02[a]

[a] Other elements like Cu, Mn, Co, etc. are even smaller.

to organic molecules [1]. Furthermore, complex processes such as ATP generation, transmembrane electric potential maintenance for electrical signal transmission, balancing the high negative charge along nucleic acid polyphosphate backbones, bone formation, and oxygen transport all require the involvement of metal ions. Among the 20 essential elements for biological systems, 10 are metal ions, comprising close to 1.5% of the body's weight. However, life depends to a far greater extent on these elements than this figure indicates. The selection of certain elements as essential raises intriguing questions. It will be very interesting and important to understand the criteria for the selection of essential elements. However, prior to that, it will be important to classify the elements based on their biological roles.

CLASSIFICATION OF ELEMENTS ACCORDING TO THEIR BIOLOGICAL FUNCTION

As described in the previous section, there are around 25 elements in the Periodic Table that are considered essential for survival. The remaining elements can be further classified into different categories based on their biological role, as shown in Table 1.2 and presented in the biological Periodic Table of elements (Figure 1.2).

Essential elements are crucial for our survival. They have a specific role in maintaining the normal living state of our body. A consistent deficiency in their intake results in an impairment of a function, and dietary supplementation of that element—and only that element—prevents the adverse effect [3]. The essential elements include both metals and non-metals. It is important to note that some elements are useful only for enhancing organism functionality and are not absolutely necessary for survival. Such elements should be differentiated from essential elements and are termed beneficial elements. For example, selenium is considered beneficial for plant growth. Essential elements are further classified into bulk essential and trace essential, depending on the quantity present in the human

TABLE 1.2
Classification of Elements and in Particular Metal Ions

Classification	Brief Description	Examples
Bulk essential elements	Absolutely required in the human diet	O, C, H, N, P, Na, K, Mg, Ca, etc.
Essential trace elements	These are also essential	Fe, Cu, Mn, Zn, Co, Mo, I, Se, etc.
Non-essential elements	Do not play a role in meeting nutritional demands	Al, Sr, Ba, V, Ti, Sn etc.
Toxic	Disturb the natural function of biological systems	Cd, Hg, Pb, As, etc.
Probes or drugs	Used in diagnosis or therapeutics	Tc, Gd, Ag, Au, Pt, etc.

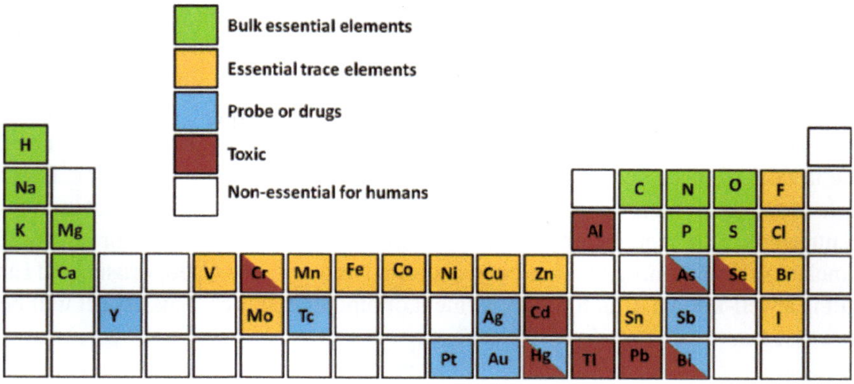

FIGURE 1.2 Classification of elements as bulk essential, trace essential, probes or drugs and toxic elements in the periodic table.

body system [4]. Bulk essential metal ions, commonly referred to as macronutrients, are metal ions required in amounts higher than 100 mg/day. They provide essential ions in body fluids and form major structural components of the body. Trace essential metal ions, commonly referred to as micronutrients, are required in small amounts (between 1 and 100 mg/day) by adults. Since it was challenging to detect such low concentrations of micronutrients, they were relatively slow to be recognized as essential elements [2]. Iron was the first to be recognized as essential when it was discovered that anemia is caused by iron deficiency and could be cured by supplementing the diet with iron-rich sources. Likewise, metals such as molybdenum, nickel and cobalt were not considered essential; only in last century has our understanding of their biological roles grown. We owe this newfound knowledge to modern analytical techniques capable of detecting elements even at the nanogram level.

Isn't it fascinating that elements present in such minuscule quantities can have such a prominent effect on an organism's health? However, our understanding of the mechanism of action of each trace element is still incomplete and it is very probable that in the future, some other elements in ultra-trace quantities (less than 1 mg/day) will be found to be essential. Unraveling these secrets is one of the major goals of modern chemical research. It is worth mentioning that trace and ultra trace essential elements are vital for an organism's functioning as long as they are within the recommended limits. If present beyond a certain limit in the human system, they can lead to damaging effects. This aspect will be elaborated further in this book.

Apart from essential metal ions, there are several other metals found in living system that do not have any specified role. These elements are considered non-essential elements. They generally exhibit close chemical resemblance (atomic size, oxidation state, etc.) to vital essential metals. Since the system for selecting metal ions is not very selective, non-essential metal ions are also being taken up by the living system along with essential ones [3]. For example, Sr^{2+} and

Ca^{2+} belong to the same group of the Periodic Table, exhibiting close similarity in chemical properties. Since Ca^{2+} is an essential component of bones, a significant amount of Sr^{2+} can be observed in bones bound to phosphate groups. Na^{+} and K^{+} are replaced by other members of their group, such as Li^{+}, Rb^{+}, and Cs^{+}. Based on a similar charge-to-size ratio, Zr^{4+} and Al^{3+} are found to replace Fe^{3+}. Thus, the uptake of metal ions by the body is not a criterion for essentiality [3]. These non-essential elements can cause adverse effects in the body even at lower concentrations. However, sometimes they are used as beneficial drugs. For instance, ranelate salt containing Sr^{2+} has been classified as a drug for osteoporosis, and Li^{+} salt is a well-known drug for the treatment of bipolar disorder. This aspect will be discussed in detail elsewhere in the book.

Among non-essential elements, there is another class called toxic elements. Non-essential elements like cadmium, mercury, and lead [5] when incorporated into the human system, cause serious problems within the body by interfering with the functions of essential metal ions. It is not that only non-essential metal ions are toxic; even essential metal ions, beyond some critical concentration, can be toxic, often in specific tissues or organs.

Beyond essential metal ions, there exists a diverse array of metal ions in the realm of modern medicine, comprising the final category in the biological classification of these elements. The utilization of metal ions in medical applications has been realized by humans since antiquity. Throughout history, metals such as gold, arsenic, copper, and iron have been used to treat different human diseases. Over time, various other metal ions have been identified for therapeutic use. For example, lithium (Li) salts are utilized to treat bipolar disorder. Platinum (Pt) and ruthenium (Ru), despite being toxic, are important therapeutic adjuvants in cancer therapy. Along with Pt, several radiolabeled complexes are quite useful in shrinking tumor cells. Gold (Au) compounds are extensively used in the management of cancer as well as rheumatoid arthritis. Besides the therapeutic role, several metal ions serve as diagnostic agents. For instance, gadolinium (Gd) complexes are increasingly used as magnetic resonance imaging (MRI) contrast agents and 99m-technetium (^{99m}Tc) is a well-known radionuclide agent approved for diagnostic imaging of abnormalities in the brain, bone, kidney, heart, thyroid, etc. From the above discussion, it can be understood that metal ions play a variety of roles, from being essential to being used in therapeutics [2].

WHY ONLY CERTAIN ELEMENTS ARE ESSENTIAL?

In the previous sections, it is mentioned that there are ~20 elements that are essential for the normal functioning of the body. Then the question arises as to why, out of the entire periodic table, only a few elements are incorporated into the living system, i.e. they have a biological role. How has nature selected certain metal ions and left others? More precisely, why are only certain inorganic elements/compounds specifically required to perform specific functions? The basic principles that govern the essentiality criteria are:

Abundance and Bioavailability

It has been observed that all the essential elements are reasonably abundant in the Earth's crust or seawater. When different metals can perform the same function, organisms tend to utilize the most abundant one [6]. Generally, lighter elements are more abundant and thus essential for biological roles. However, there is an exception to this rule. Molybdenum (Mo), which has low amounts in the Earth's crust. However, Mo as MoO_4^{2-} has as much concentration as iron in seawater. Since there is enough reason to believe that life originated in the oceans, it becomes imperative to take into account the concentration of metal ions in seawater as well. Zinc and copper share a similar case, as their concentrations are also comparable to that of iron in seawater. All these four elements are regularly found at the catalytic sites of enzymes.

Calcium compounds are used to perform structural roles, unlike strontium compounds, and the reason is obvious: calcium is more abundant than strontium. The most abundant alkali metal ions (sodium and potassium) are selected by nature to transmit signals and control ion balance. Another interesting fact is that zinc and cobalt are easily interchangeable, and *in vitro* Co(II) can easily replace Zn(II) in zinc enzymes without affecting catalytic activity. Nonetheless, organisms choose zinc as it is more abundant in both the Earth's crust and seawater.

Exceptions to the rule as mentioned earlier are aluminum (8.2%), silicon (28.2%), and titanium (0.57%), which, although abundant in the Earth's crust, are not essential elements. They play a negligible role in biological systems because their compounds, oxides, or hydroxides are insoluble in water at physiological pH and thus play a negligible role [7]. Conversely, less abundant elements like molybdenum have high water solubility in the form of MoO_4^{2-} at physiological pH, and it is essential in many living organisms. This indicates that natural abundance is not the sole criterion. Apart from this, the metal ions should be bioavailable, i.e., in the right place at the right time and readily assimilated by the biological system.

Ability, Efficiency and Specificity

In addition to abundance and bioavailability, another critical factor for an element to be essential is its ability to perform vital functions crucial for life. If two or more moieties can perform the same function, nature selects the most efficient one [6]. For instance, flavodoxins and ferredoxins transport electrons in similar ways, but ferredoxins, being more efficient, are the prime choice of organisms.

Often, biological ligands also fine-tune the properties of the metal center to perform specific functions. In enzymatic reactions, the metal center should be in a distorted geometric environment where its activity becomes very specific, and at the same time, the activation barrier for the reaction is reduced [6]. These aspects are mainly controlled by the protein portion of the metalloproteins. A closer look at the active site of these metalloproteins reveals that the metal

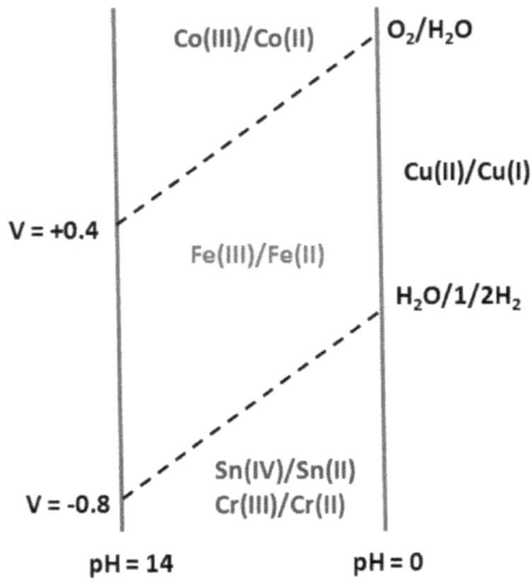

FIGURE 1.3 Schematic representation of the reduction potential of some metal systems. Redox systems within the potential range marked by dotted lines are suitable for performing redox functions in biological systems.

present there have distorted coordination environment, which is probably responsible for their high catalytic activity. Structural studies performed on blue copper proteins and carboxypeptidase confirm the distorted tetrahedral geometry around copper and zinc in these metalloproteins [8].

For metal ions to perform redox functions in biological systems, the primary requirement is that the metal ion should not decompose water (Figure 1.3), i.e., they should not oxidize or reduce water. This may be why certain redox catalysts, such as Co(III)/Co(II), Sn(IV)/Sn(II), and Cr(III)/Cr(II), which have redox potentials outside the decomposition limit of water, are excluded from performing redox activity in biological media [6]. Conversely, iron-containing metalloenzymes are confined within the range of the reduction and oxidation potential of water. This could be one of the reasons, along with the abundant nature of iron, why iron-based catalysts are ubiquitous in biological systems. Another interesting point to note in this sequence is the reduction potentials of O_2^-/H_2O_2 and O_2/O_2^-, which are +0.96 V and −0.45 V (at pH 7), respectively. Redox systems within this potential range can function as catalysts for the superoxide dismutase (SOD) reactions. Cu(II)/Cu(I) is a qualified redox system that falls within this range and is indeed found in natural SOD [6]. Other redox potential that fall within this range are Fe(III)/Fe(II) and Mn(III)/Mn(II). These redox systems are utilized to synthesize SOD mimics.

Discussion over cobalamin, commonly known as vitamin B12, is another intriguing example of how the ability to perform particular functions qualifies an element as essential. Cobalamin-dependent enzymes are involved in various critical processes such as methionine synthesis, isomerization reactions, dehalogenation reactions, etc. For a molecule to perform the functions of cobalamin, it should readily exhibit three consecutive oxidation states, such as +I, +II, and +III, in aqueous media. Additionally, the lowest oxidation state should be highly nucleophilic and the middle oxidation state must have an unpaired electron [9]. Only transition metals in the periodic table can meet the first criterion of exhibiting three consecutive oxidation states. Meeting the second criterion, where the metal ion should be highly nucleophilic in the lowest oxidation state, requires a d^8 or d^{10} configuration. This criterion could be met by Fe(0), Co(I), Ni(0), and Cu(I). However, Fe(0) and Ni(0) are not stable oxidation states, while Cu(I) is not highly nucleophilic. Moreover, Cu(III) is not stable at all under normal conditions. The only remaining option is Co(I), which uniquely satisfies the requirements [9].

Kinetic and Thermodynamic Factor

Nature utilizes those metal ions that form kinetically labile and thermodynamically stable metallobiomolecules with biological ligands. Thermodynamic stability ensures that metallobiomolecules have sufficient life in the biological environment to perform their function, whereas kinetic lability promotes the fast assembly and disassembly of the metal cores as well as the rapid association and dissociation of substrates. Inert metal complexes lead to accumulation of metal ions in the biological system, which may result in the toxicity of those metal ions. For this reason, Cr^{3+}/Co^{3+} and transition metal ions belonging to the 4d and 5d series are rarely utilized in biological systems [7].

At this point, it is worth mentioning that when metal ions required for a specific purpose are not abundant in the Earth's crust or seawater, these metal ions can be concentrated in the body by energy-driven processes. Through these processes, which involve ATP, the metal ion is concentrated to several orders of magnitude in the organism compared to its concentration in seawater. The uptake of vanadium (V) in some sea squirts falls into this category [7].

BIOLOGICAL LIGANDS

Having established the selection criteria for metal ions in biological roles, our focus now shifts toward comprehending the intricate processes by which these metal ions are inserted into the biological environment from their mineral forms and subsequently absorbed by the cells. To perform biological role, a metal ion must first dissolve in the biological fluid, whose pH is in the range of 6–8. While metal ions such as sodium, magnesium, zinc, etc., which are readily soluble under physiological pH conditions, present no significant challenges, However, other metal ions, such as iron, which predominantly exist as insoluble ferric hydroxide in this pH range. Therefore, despite being abundant, iron is not readily available to cells under such circumstances. To address this issue, organisms have developed specialized mechanisms to overcome the limitation and facilitate the solubilization and mobilization of such metal ions [7].

To unravel this process, attention is directed toward the strategy employed by bacteria to solubilize and mobilize iron in biological fluid. The underlying principles apply to other metal ions as well. Bacteria secrete powerful and selective chelating agents called siderophores. These siderophores display a very high affinity for iron (III) and form stable complexes, thereby effectively solubilizing the otherwise insoluble iron [10]. Importantly, these siderophores are secreted specifically in response to iron deficiency. Upon secretion these siderophores engage in complexation with iron, facilitating its uptake into the cell. One such siderophore is the catechol-based chelating ligand enterobactin.

The structure of enterobactin, depicted in Figure 1.4, reveals an intriguing arrangement wherein three catechol rings are attached to each other via a rigid twelve-membered ring of carbon and oxygen. Within this structure, hydrogen bonding interactions occur between the amide –NH and O atom of the catechol ring, thereby restricting the free rotation of the catechol groups and forcing them to face toward the center of the ligand [11]. Spectroscopic studies revealed that the iron (III) coordinates to the catecholate dianions formed after deprotonation rather than through amide oxygens or ring oxygens. This observation highlights a common phenomenon in biological coordination chemistry known as ligand preorganization, wherein the ligand's functional groups exhibit a specific arrangement

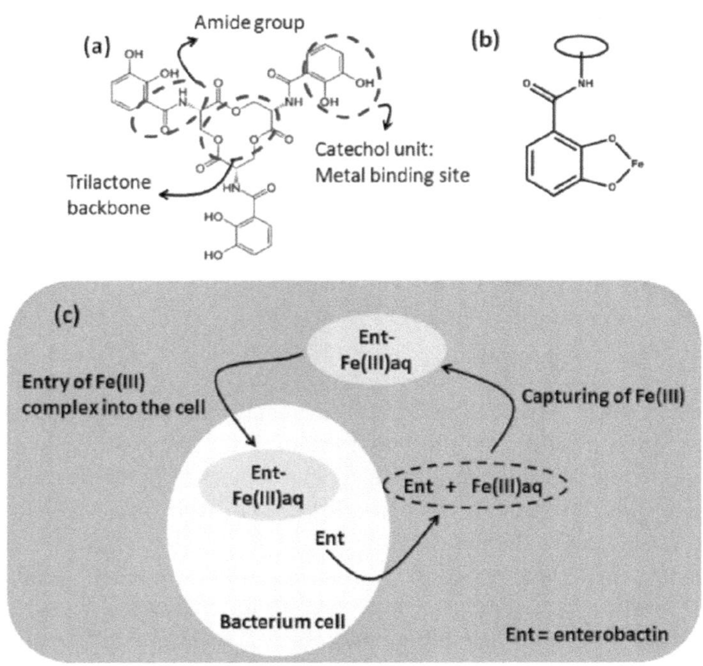

FIGURE 1.4 (a) Chemical structure of enterobactin displaying various functional units, (b) representative binding mode of iron (III) with the catecholate functional group, and (c) schematic representation of the uptake of iron by enterobactin secreted by bacterial cell.

prior to the metalation process. This preorganization is attributed to providing selectivity to the ligand and enhancing the stability of the resulting complex.

The stereochemistry of enterobactin plays an important role in the recognition, encapsulation, and transport of iron. The Fe-enterobactin complex binds with specific receptors located on the outer membrane of bacteria (E. coli) during iron stress conditions. This complex is subsequently internalized into the cell through an active transport mechanism. Studies suggest that the tris(catecholato) and amide portions of the Fe-enterobactin complex are essential for its recognition and interaction with the membrane receptor [12].

Subsequent to the absorption of the Fe-enterobactin complex by the E. coli cell, the release mechanism of iron from the iron-enterobactin complex is equally intriguing. The central trilactone ring of enterobactin serves as the primary factor contributing to the formation of such a stable complex with iron. If the structure of this ring is destroyed, enterobactin will no longer be able to hold iron effectively. The loss of chelating power of enterobactin is attributed to the release of iron. Esterase enzymes hydrolyze the trilactone ring, which disrupts its structure and thereby reduces the hold of enterobactin over iron. The loosely bound iron binds with specific intracellular ligands which causes the reduction of Fe(III) to Fe(II), forming a more labile and less stable form. This aids the release of iron from enterobactin in cells [11]. A schematic representation of iron uptake and release by enterobactin is depicted in Figure 1.4C.

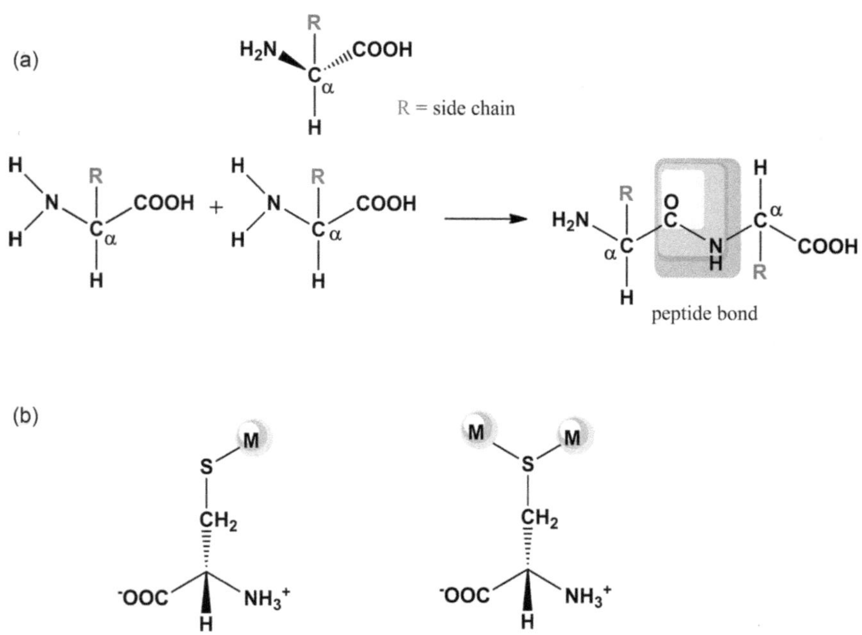

FIGURE 1.5 (a) Representative structure of α-amino acids displaying the amino group, carboxylic group, and side chain, (b) schematic representation of the formation of a peptide bond.

Thus far, it has been established that the solubilization of metal ions in biological fluids and their subsequent transportation and absorption by cells are crucial processes for their utilization in biological functions. For the latter part, two possible pathways have been identified: (i) ions such as sodium are transported inside the cells through specialized pumps and channels present in the cell membrane, and (ii) there are special chelating agents for the rest of the ions that assist their transport across the cell membrane [7]. To explain these phenomena, we have considered the case of iron; however, it is not only one element that is absorbed by the cell. Several elements are absorbed simultaneously, and a situation arises where there can be synergism (positive interaction) as well as antagonism (negative interaction) between the ions, i.e., the interactions of an ion with another ion will either facilitate or inhibit its absorption. For instance, cadmium uptake by the cells inhibits the absorption of iron and copper, while zinc uptake doesn't interfere with the absorption of copper; however, it affects iron uptake.

Once metal ions enter the biological system, they are encountered by the ligands present in that environment, which tune their properties to perform specific functions [13]. The ligands present in the biological environment can be classified into the following categories:

a. Amino acid side chains
b. Constituents of nucleic acid
c. Organic cofactors
d. Inorganic anions like phosphate, bicarbonate, etc.

Amino acid side chains as ligands: In our biological system, there are 20 alpha (α)-amino acids used in the synthesis of proteins. α-amino acids are those amino acids in which the amine and carboxylic acid parts are bound to the same carbon called α carbon. A representative structure of an α-amino acid is shown in Figure 1.5a.

Proteins consist of unbranched α-amino acids joined by the peptide bond shown in Figure 1.5b. Since the amino and carboxylic acid parts of amino acids are involved in the formation of peptide bonds, they mostly function as ligands through their side chains. Depending on the side chains, the 20 naturally occurring amino acids can be classified as nonpolar, polar neutral, and polar charged, as shown in Figure 1.6.

Among the diverse array of naturally occurring amino acids, certain side-chain ligands have been observed to play prominent roles in biological systems. Cysteine's thiolate and tyrosine's phenolate group, active around pH 10 are commonly encountered ligands. Histidine's imidazole group is available for coordination around pH 6 while carboxylates of glutamic and aspartic acids, accessible even at the lower pH range of 3–4, are also frequently observed ligands in biological systems. Except for tyrosine, each of these side-chain ligands can coordinate in terminal as well as bridging modes [14]. Possible interaction modes of the cysteine side chain are displayed in Figure 1.7. The hydroxyl groups of serine and threonine, which deprotonate around pH 13, the thioether group of methionine,

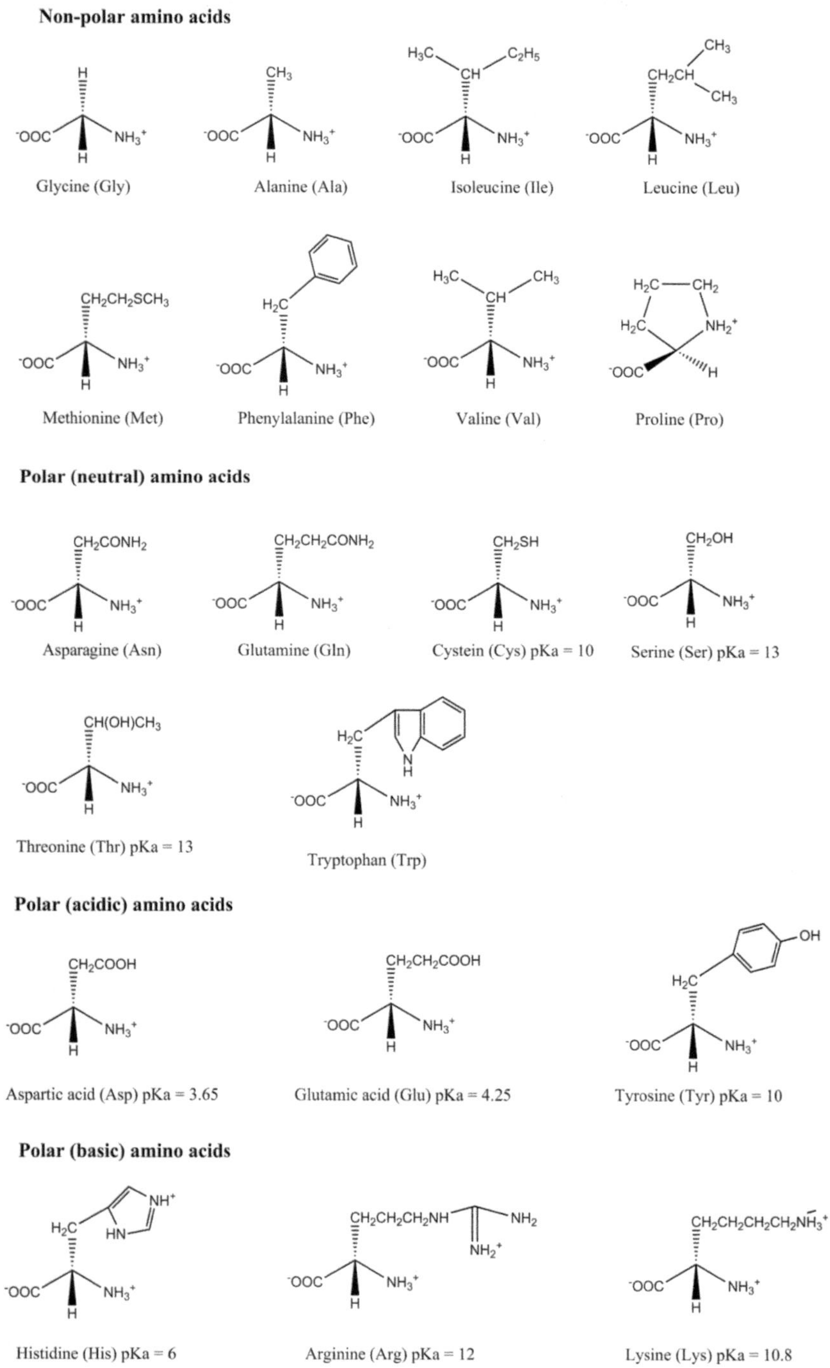

FIGURE 1.6 The figure depicts the chemical structures of the naturally occurring amino acids classified based on their polarity.

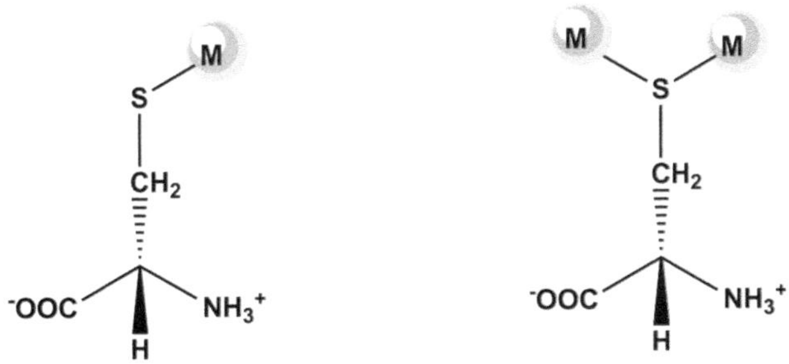

FIGURE 1.7 Schematic representation depicting the possible binding modes, i.e., linear as well as bridging, exhibited by the thiolate of the cysteine residue.

the amino group of lysine, the guanidine group of arginine, and the carboxamide groups of asparagine and glutamine are also potential coordinating sites, though less commonly encountered compared to cysteine, histidine, tyrosine, glutamic acid, and aspartic acid.

In addition to amino acid side chains, other peptide components such as the peptide carbonyl group, deprotonated atoms of the peptide bond and the amino and carboxylic groups of N-terminal and C-terminal amino acids, respectively, also serve as ligands for coordinating with metal ions. It is worth emphasizing once again that these α-amino acids are fundamental units in protein synthesis in biological systems and important factors in determining the structure of proteins. Protein structure can be analyzed at four different levels. The primary structure of a protein refers to the linear sequence of amino acids connected through covalent bonds to form a polypeptide chain. The structure and function of proteins critically depend on the specific composition and sequence of these amino acids. Secondary protein structure, encompassing α-helices and antiparallel or parallel β-pleated sheets, arises from the local folding of the polypeptide chain driven by intermolecular interactions, particularly hydrogen bonding. Tertiary protein structure describes the overall three-dimensional arrangement of the polypeptide chain. The folding of the polypeptide chain into a unique three-dimensional structure is influenced by various forces and interactions, including hydrophilic and ionic interactions on the protein surface, disulfide bonds as covalent cross-links, hydrophobic interactions among nonpolar amino acids, and coordination of amino acid side chains with metal ions. Finally, the quaternary structure of a protein refers to the association of multiple polypeptide chains in a tightly packed structure through hydrogen bonds and van der Waals interactions. The availability and orientation of amino acid side-chain ligands in a metalloprotein greatly depend on the overall three-dimensional structure of the protein [14].

There are numerous examples in biological systems where a metal ion is coordinated to amino acid side chains. One of them is the iron transport protein in the human body, called transferrin. We have already seen that bacteria secrete siderophores (e.g., enterobactin) to capture iron(III). Therefore, iron in the human body must be kept under strict watch, as any free iron within the human system is likely to be chelated by siderophores, which may lead to bacterial infection, precipitate as iron(III) oxide, or form iron(II). The formation of iron(II) species leads to the production of hydroxyl radicals, which pose significant danger to living cells. Concerns about infection, toxicity, and insolubility have led to the evolution of transferrin protein for acquiring and transporting iron. In transferrin, iron(III) is coordinated by one carboxylate oxygen from aspartic acid, one imidazole nitrogen from histidine, and two phenolate oxygens from tyrosine side chains. In addition, a carbonate ligand is also held in place within the protein via hydrogen bonds to the side chains of arginine, threonine, and two peptide—NH groups present in the protein backbone [15]. This results in ligand preorganization, creating an octahedral environment to capture the iron(III) ion, as shown in Figure 1.8.

Nucleic Acid and Their Constituents

Nucleic acids are naturally occurring chemical species that play a crucial role in storing and expressing genetic information. They consist of three main components: a sugar, a phosphate group, and a nitrogenous base. Depending on the type of sugar and base, nucleic acids are classified into two classes: deoxyribonucleic acid (DNA) and ribonucleic acid (RNA). RNA contains ribose sugar, which has a hydroxyl group at the C-2 position, while DNA has deoxyribose sugar, which lacks the hydroxyl group at this position. The nitrogenous bases are divided into two

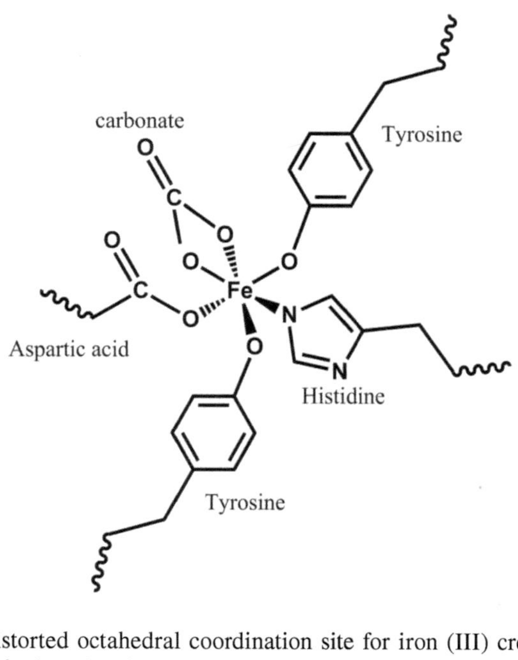

FIGURE 1.8 Distorted octahedral coordination site for iron (III) created by aspartyl, histidyl, two tyrosinyl, and carbonate ligands, held in place by hydrogen bonds to amino acid side chains inside the small cleft.

categories: purines and pyrimidines. Purine bases include adenine and guanine while pyrimidine bases comprise cytosine and thymine (in DNA) and uracil (in RNA). Each sugar molecule is connected to one of these five nitrogenous bases, forming a nucleoside. A nucleoside along with a phosphate group linked to the C-3 or C-5 position of the sugar is referred to as a nucleotide. Adjacent nucleotide residues are linked together by phosphodiester bonds.

The DNA or RNA molecule can adopt a wide range of conformations, which are determined by the torsion angles in the phosphodiester bonds and sugar rings. Additionally, the bases can take on either a *syn* or *anti* conformation with respect to the bond linking the sugar and the base, leading to different overall conformations. These variations can have profound effects on the tertiary structure of nucleic acids and, in turn, their metal-binding characteristics.

The primary metal-binding sites in nucleic acids are the nitrogen and oxygen atoms of nitrogenous bases [14], as highlighted in Figure 1.9. However, since the lone pair of electrons residing on the nitrogen atoms present in adenine, cytosine, and guanine are delocalized into the ring by resonance, it is highly unlikely that these molecules coordinate with metal ions at neutral pH. These species usually interact with metal ions at high pH, where deprotonation results in a nucleophilic RNH– ion. Negatively charged oxygen atoms of the phosphodiester bond mark another binding site in nucleic acids [14]. It is important to note that, whereas nitrogenous bases mostly serve as binding sites for transition metal ions like Cu and Pt, the phosphodiester bond binds with hard metals like alkali and alkaline earth metal ions. For instance, the binding of Mg^{2+} by the oxygen of the phosphodiester bond helps stabilize the tertiary structure of RNA.

Special Ligands

Properties of metal ions are tuned by their local environment to perform their roles in biological systems. Sometimes, however the amino acid and nucleic acid residues are unable to tune the properties of metal ions alone. Nature has addressed this issue by synthesizing specific organic functionalities that are capable of modulating the properties of metal ions in conjugation with proteins. These special ligands are also called prosthetic groups. The chemical structures of some prosthetic groups are shown in Figure 1.10.

Within the realm of prosthetic groups, the porphyrin molecule has emerged as one of the most extensively utilized moieties. These captivating structures consist of four pyrrole (five-membered heterocyclic rings) units linked in a cyclic manner. The iron-porphyrin complex, known as heme, has been the subject of intense scrutiny due to its pivotal role in numerous essential functions ranging from oxygen transport and storage to oxygen activation and electron transport. This showcases the versatility and indispensability of heme in biological processes. Another interesting class of porphyrin complexes comes in the form of magnesium-porphyrins, famously recognized as chlorophylls. These molecules are essential for photosynthesis, harnessing the radiant energy of the sun and

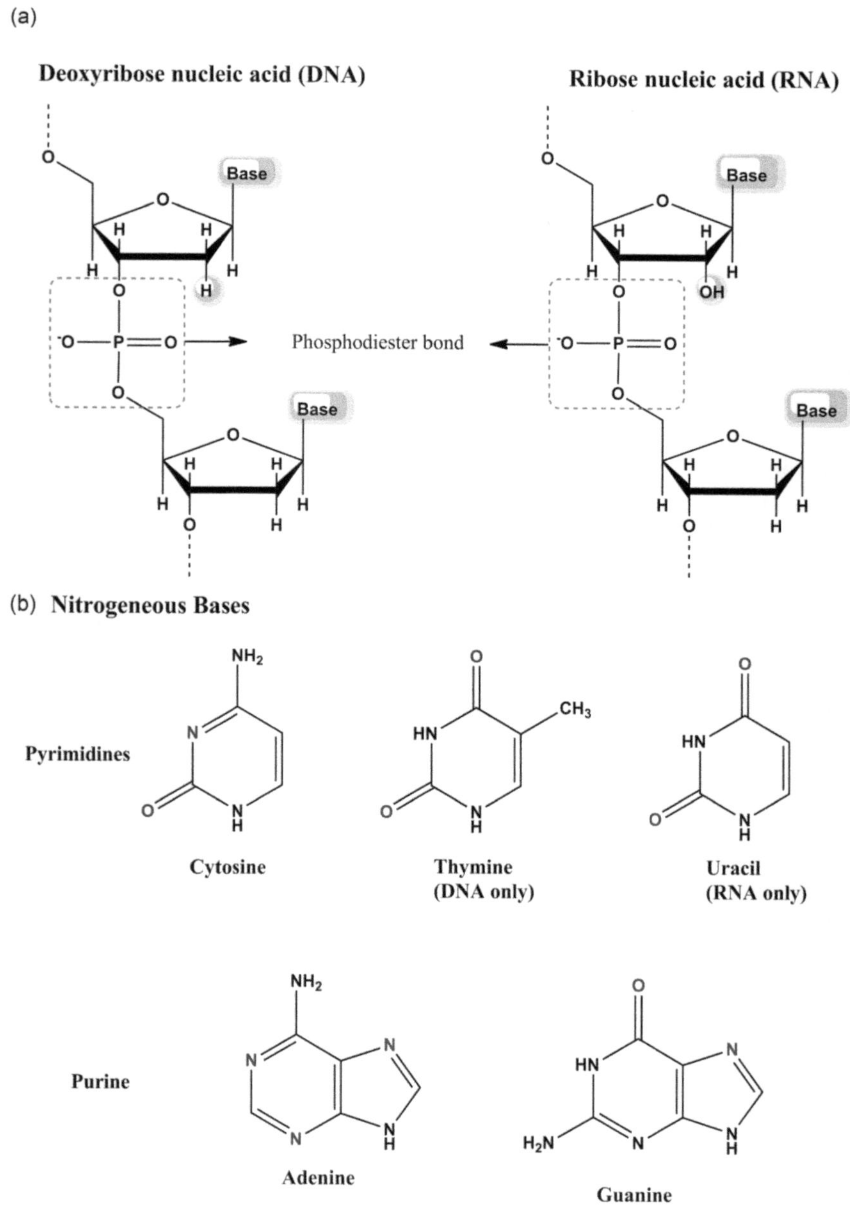

FIGURE 1.9 (a) Structure of DNA and RNA. The basic differences between the two structures. (b) The chemical structure of nitrogenous bases present in DNA and RNA.

converting it into chemical energy. At this juncture, it becomes crucial, from a biochemical perspective, to unravel the mysteries surrounding the origin of these specialized ligands within our biological system. The biosynthesis of porphyrin starts with the formation of 5-aminolaevulinate (ALA) as a precursor molecule. In plants, ALA is synthesized by the reduction of glutamate to glutamate-1-semi-aldehyde, whereas in higher organisms other than plants, ALA is synthesized by the condensation of succinyl-CoA, derived from the citric acid cycle, and glycine. Subsequently, two ALA molecules condense to form pyrrole. Four pyrrole molecules then undergo oligomerization to form a tetrapyrrole linear chain, followed by ring closure. The planar porphyrin ring is unable to chelate to a metal ion. Distortion of the ring, wherein two opposite pyrrole rings move above the plane while the remaining two rings move below the plane, generates an appropriate configuration suitable for metalation. The deprotonated nitrogen of the pyrrole rings coordinate first, followed by the sequential deprotonation and coordination of the remaining two nitrogens of the pyrrole rings [13].

Corrin rings, closely resembling porphyrin rings but lacking a methine carbon, offer another intriguing example within this class of ligands. This subtle structural variation around the metal center introduces distortions and contributes to the remarkable properties observed in the resulting complexes. Notably, cobalt forms the primary metal-corrinoid complex in Vitamin B12, while nickel-corrin complexes occur naturally and play a crucial role in hydrogenase enzymes.

Beyond porphyrins and corrins, specialized ligands such as hydroporphyrins, specifically the reduced form of porphyrin known as Factor 430, exhibit distinct characteristics. Nickel-hydroporphyrin complexes function as prosthetic groups in S-methyl coenzyme M reductase, showcasing the diversity and importance of these ligands in various biological processes [14].

Inorganic Anions

In addition to amino acid residues, nucleic acid constituents, and specialized ligands, there exists a diverse array of low molecular weight inorganic ligands that play a crucial role in metal binding within metalloproteins. These include carbonates (CO_3^{2-}), bicarbonates (HCO_3^-), phosphates (PO_4^{3-}), cyanides (CN^-), carbon monoxide (CO), etc. [13]. These inorganic ligands contribute to the intricate metal coordination networks observed in various biological systems.

A suitable example of this class is transferrin, where the carbonate ion engages in bonding with iron (III), alongside histidine, aspartic acid, and tyrosine molecules. Similarly, cyanide (CN^-) and carbon monoxide (CO) serve as ligands to iron in bacterial hydrogenase. Phosphates (PO_4^{3-}) or bicarbonates (HCO^{3-}) function as ligands in ferric ion-binding proteins found in pathogenic bacteria such as Neisseria and Haemophilus. Interestingly, these ferric ion-binding proteins employ a remarkable 'Venus flytrap' mechanism to capture iron directly from iron transport proteins like transferrin in mammals [13].

In addition to the aforementioned classes of biological ligands, intriguing and intricate metal-binding assemblies arise from the collective arrangement of various biomolecules. These include cell membranes and intracellular machinery like ribosomes, mitochondria, the endoplasmic reticulum, etc. These complex structures serve vital functions, with cytochromes embedded in the cell membrane

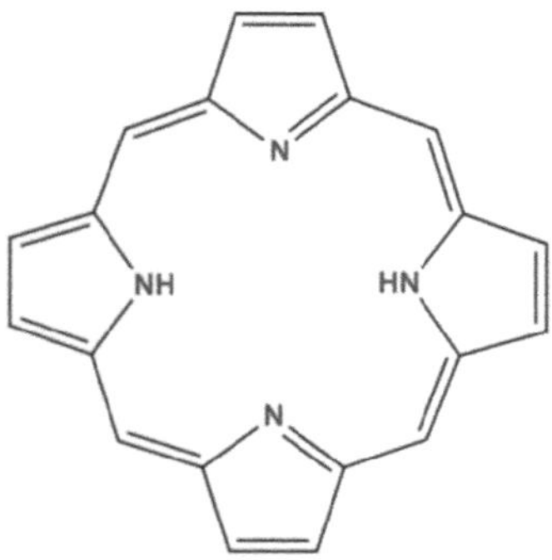

Porphyrin

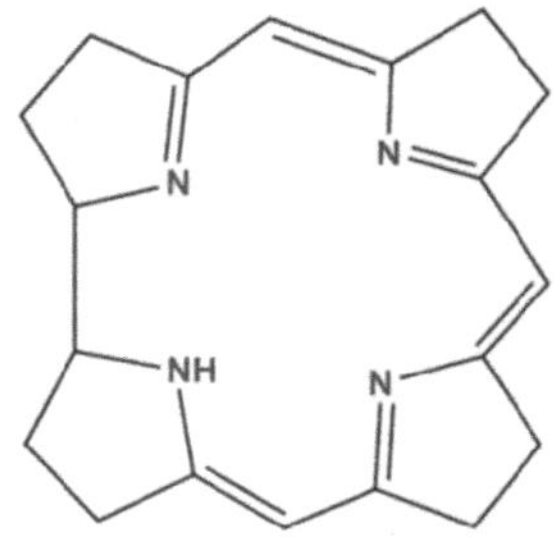

Corrin

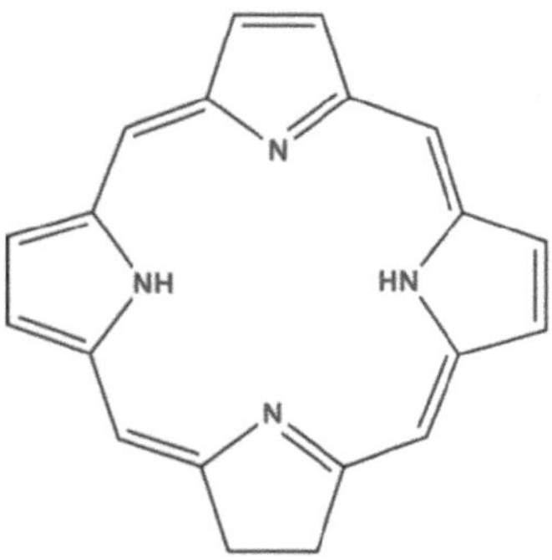

Chlorin

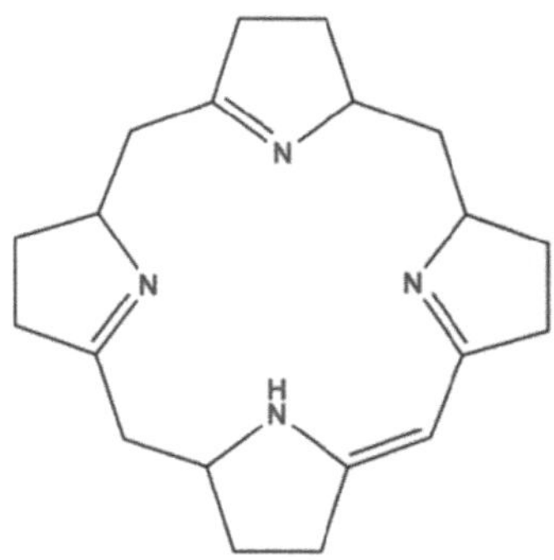

Factor 430

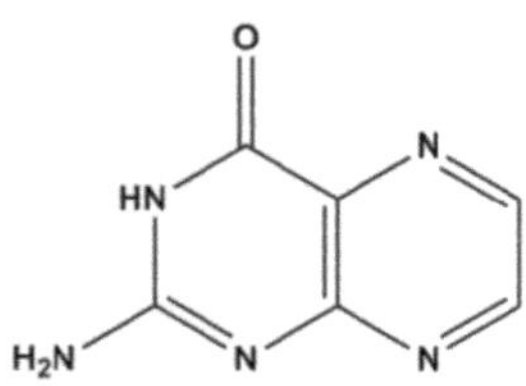

Pterin

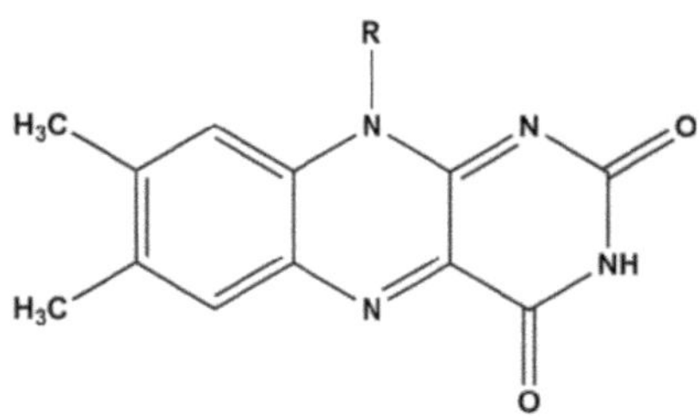

Flavin

FIGURE 1.10 Chemical structure of some of the representative ligands.

facilitating electron transfer and oxidative phosphorylation, and other biological structures facilitating metal ion transport across the cell membrane, among other remarkable processes.

It is within this intricate interplay of ligands and metal ions in the biological environment that metallobiomolecules emerge, playing a crucial role in body functioning.

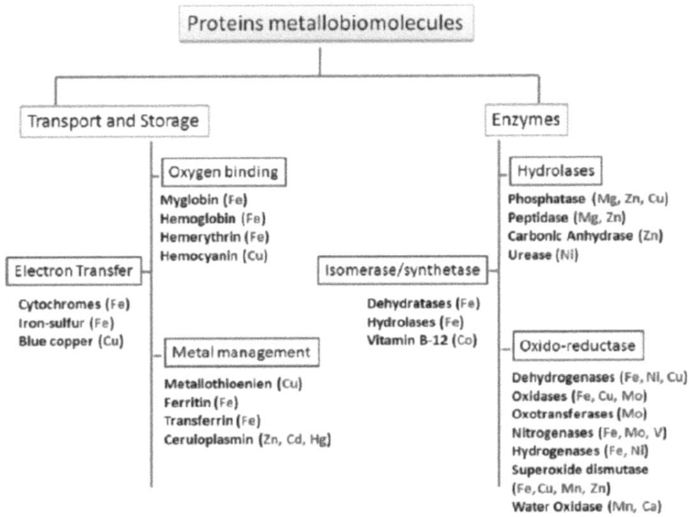

FIGURE 1.11 Classification of metallobiomolecules. (Copyright permission from Elsevier, Ref. [16].)

METALLOBIOMOLECULES

Metallobiomolecules are essential for regulating biological functions. They consist of two components: the molecular part and the metal ion. The molecular part refers to the ligands present in biological systems. These ligands coordinate with the metal ion under the biological environment to form metallobiomolecules.

Depending on the molecular part, metallobiomolecules can be classified as non-protein metallobiomolecules and protein metallobiomolecules, as shown in Figure 1.11. The non-protein metallobiomolecules are responsible for carrying out two primary functions: photoredox function and metal ion storage and transportation.

Photoredox Function of Non-Protein Metallobiomolecules

Energy from the sun enters the Earth's atmosphere through the process of photosynthesis, which is considered an oxidation-reduction reaction induced by light. This results in the evolution of molecular oxygen from water and the formation of carbohydrates from the reduction of carbon dioxide. Chlorophyll, a magnesium-porphyrin non-protein biomolecule, is present in plants and plays a crucial role in photosynthesis. It involves the absorption of solar energy and utilization in the photolysis of water molecules to replenish the reducing power of the cells [17]. It is believed that materials capable of mimicking the function of chlorophyll can convert CO_2 into useful chemicals like methanol, formic acid, and formaldehyde, thereby helping to reduce CO_2 levels in the air and provide solar fuels simultaneously.

Metal Ion Transport and Storage Function of Non-Protein Metallobiomolecules

This class of molecules includes siderophores responsible for the transport of iron in bacteria, calcium phosphate embedded in a fibrous matrix, which is a major constituent of bones, and silicon in the form of silica, used for structural purposes in diatoms, some protozoa, and sponges. Inspired by the functions of such metallobiomolecules, many synthetic compounds are being explored as artificial organs. For instance, silicones, used as artificial skin, are gaining paramount importance in the field of medical sciences.

Protein metallobiomolecules, alternatively called metalloproteins, are a class of compounds that contain a protein-bound metal site. This site consists of one or more metal ions coordinated with amino acid side chains and other biological ligands. How the amino acid side chain interact with the metal has already been discussed in the previous section. Metalloproteins account for nearly half of the proteins reported to date and are essential for proper physiological function [18]. Metal ions play a crucial role in governing the function of metalloproteins. They are mainly involved in catalysis, metal-oxygen coordination and de-coordination, the uptake, release, and storage of electrons, and stabilizing the tertiary and quaternary structures of proteins [19]. Furthermore, the metal site in metalloproteins also acts as a site for the uptake, binding, and release of metals in soluble form.

The intricate ensemble of proteins and biological ligands in metalloproteins empowers the metal ions to fulfill their functions with high efficiency. These metal ions are primarily enclosed within macrocyclic ligands, such as porphyrin and corrin, as illustrated in Figure 1.10. These ligands have a remarkable influence on the properties of the metal because of the delocalization of metal d-orbitals into the π-orbital network of the ring. Notably, the iron residing within the heme environment manifests striking dissimilarities when compared to its non-heme counterparts. Additionally, the structure of the protein in metalloproteins significantly impacts the electronic structure, redox potential, and detailed arrangement of the metal center. The protein structure also plays an important role in facilitating allosteric interactions and cooperative binding, organizing catalytic and electron transfer sites. Moreover, proteins can provide surface recognition sites for strong interactions with complementary regions, as well as channels and pathways for the metal center to access its active site. Protein structure can also create a hydrophobic environment that helps stabilize bound molecules, like dioxygen in the case of hemoglobin. Furthermore, the protein structure can force the metal to adopt unusual geometries, which are high-energy states. This arrangement enhances the reactivity of the metal center, aiding in processes such as electron transfer and catalysis [19].

Empowered by these important characteristics, metalloproteins play vital roles in biological systems, fulfilling various functions. Some important functions performed by metalloproteins are briefly discussed in the next section.

Metalloproteins such as hemoglobin, myoglobin, hemocyanins, and hemerythrins are uniquely involved in dioxygen (O_2) binding and transport [20].

Although they possess different functional metallic cores, they all serve the common function of dioxygen transport in various organisms. These metalloproteins have a unique ability to bind dioxygen molecules without undergoing irreversible electron transfer or oxidation reactions. For example, hemoglobin and myoglobin utilize iron-porphyrin complexes that undergo structural changes upon oxygen binding, inducing cooperative movements in protein chains to efficiently uptake oxygen. In contrast, hemocyanin and hemerythrin employ paired metal ions for oxygen binding.

Another significant class of metalloproteins participates in electron transfer reactions. These reactions involve the transfer of electrons, enabling energy conversion and redox balance within the cells. Metal ions, coordinated by specifically designed ligands in metalloproteins, serve as electron carriers and catalysts in these processes. Notable examples include cytochromes, heme-containing proteins with iron ions that participate in redox reactions. Cytochromes form electron transport chains in cell membranes, facilitating electron transfer and ATP synthesis. Iron-sulfur (Fe-S) proteins are another important example, which play a crucial role in various electron transfer reactions, including photosynthesis, respiration, and DNA repair. Additionally, blue copper proteins are commonly encountered electron-transfer molecules in bioinorganic chemistry [21].

Metalloproteins also contribute to structural functions within biological systems. Metal ions provide coordination sites for ligands, thereby helping to maintain the three-dimensional conformation of proteins. Notable examples include zinc finger proteins, wherein zinc ions are coordinated by cysteine and histidine residues to form a finger-like projection that interacts with DNA and several proteins [22]. These zinc fingers play a remarkable role in regulating gene expression. Calcium-binding proteins are another remarkable example of metalloproteins involved in maintaining the structural integrity of bones, teeth, and connective tissues. Moreover, calcium-binding proteins are also involved in cellular signaling pathways [23]. Metalloproteins with structural functions are crucial for maintaining cellular architecture, supporting tissue integrity, and enabling various biological processes.

In the intricate world of biological systems, metalloenzymes represent a remarkable subclass of metalloproteins, playing a pivotal role in catalyzing a wide range of biochemical reactions. They heavily rely on metal cofactors, which are crucial for their enzymatic activity. For example, superoxide dismutase acts as a defense mechanism against reactive oxygen species (ROS) by converting superoxide radicals (O_2^-) into molecular oxygen and hydrogen peroxide. Another important class of metalloenzymes includes oxidases and oxygenases, which facilitate oxidation reactions by utilizing O_2 as a substrate. These enzymes often contain metal cofactors such as iron, copper, or manganese, which activate O_2 and enable subsequent oxidative reactions. Notable examples of this class are cytochrome C oxidase and methane monooxygenase [24]. Hydrogenases and nitrogenases are additional metalloenzymes involved in hydrogen metabolism and biological nitrogen fixation. Hydrogenase utilizes iron and nickel as cofactors, while nitrogenase employs iron with molybdenum or vanadium [18]. Hydrolases, also known as hydrolytic metalloenzymes, catalyze the cleavage of chemical bonds through the addition of water. Examples of this class include carboxypeptidase, carbonic anhydrase, and metallophosphatases, which respectively hydrolyze peptide bonds, convert carbon dioxide and water into carbonic acid, and catalyze the hydrolysis of phosphate ester bonds. Commonly employed metal ions in hydrolases include zinc, calcium, manganese, magnesium, and iron. Metalloproteins also serve as multi-electron redox enzymes involved in complex electron transfer reactions [25]. They participate in various redox processes, such as electron transport chains, metabolic pathways, and detoxification mechanisms. Notable examples of this class include cytochrome P450, which contains heme iron and catalyzes the oxidation of a wide range of substances, and iron-sulfur proteins, which contribute to electron transfer processes during energy metabolism, DNA repair, and biomolecule synthesis.

Metalloproteins are also instrumental in facilitating isomerization reactions. Cobalamin, or vitamin B12 coenzyme, is an important example of this class and plays a crucial role in catalyzing 1,2-carbon shift rearrangement reactions essential for biological processes [26]. Vitamin B12 coenzymes also catalyze methyl group transfer and dehalogenation reactions. Aconitase is another noteworthy metalloprotein involved in the conversion of citrate to isocitrate in the Krebs cycle, with iron serving as an important activator and reducing agent for this enzyme [25].

In addition to their functional roles in metalloproteins, metal ions have diverse and crucial functions in biological systems. They are actively involved in signal transduction and cellular signaling pathways. Alkali and alkaline earth ions, such as Na^+, K^+, and Ca^{2+}, act as triggers for cellular responses. The influx of Na^+ across the cell membrane and the regulation of intracellular functions by calmodulin, a calcium-binding protein, exemplify their involvement in neuron signaling and intracellular regulation [27]. Zinc finger proteins, which regulate transcription factors, also play a key role in metal ion-mediated communication. The important functions of various metal ions and their associated deficiency symptoms are listed in Table 1.3.

METAL DEFICIENCY AND TOXICITY

The preceding discussion clearly emphasizes the essential role of metal ions in various critical functions within humans and other organisms. The scarcity of essential metal ions has detrimental effects on the overall health. For instance, a deficiency of cobalt and iron may cause anemia, zinc deficiency can lead to growth retardation, and copper deficiency can result in heart disease in infants. However, it is important to recognize that essential and non-essential metal ions can become toxic when their concentrations exceed certain thresholds. As the famous saying by Paracelsus goes, 'Everything is poisonous and nothing is not poisonous; only the dose determines whether something is poisonous or not'. This concept holds true for metal ion toxicity as well. Maintaining the concentration of essential metal ions within appropriate limits in each cell and tissue is crucial

TABLE 1.3
Role of Selected Metal Ions and Their Deficiency Symptoms in Living Systems

Metal Ion	Biological Role	Deficiency Symptoms
Na and K ion	Charge carriers, membrane potential, osmotic pressure, signaling	Muscle weakness, spasms, cramps, confusion
Mg and Ca ion	Mainly structural role	Retarded skeletal growth and muscle cramps
V, Fe, Mo ion	Nitrogen fixation	Retarded cellular growth
Fe and Cu	Redox reaction, electron transfer, transport, and storage of O_2	Anemia, immune system disorder, liver disorder
Co	Oxidase, alkyl group transfer	Pernicious anemia
Ni	Hydrolase, hydrogenase	Growth depression
Zn	Structural stability and hydrolase	Skin damage, retarded growth, and sexual maturation

for proper physiological functioning, a state known as homeostasis, as defined by Claude Bernard. Any disruption in metal ion homeostasis can lead to deleterious effects on health. Bertrand's plot, shown in Figure 1.12, illustrates the relationship between metal ion concentration and physiological response.

From the plot, it is evident that there is an optimum concentration range where the physiological response is optimal for maintaining good health. However, the concentration of metal ions cannot be increased indefinitely without detrimental effects. Beyond the optimal concentration range, further increases in metal ion concentration result in diminishing physiological responses and can even be fatal, indicating the onset of the toxicity region [3]. Metal ion toxicity typically arises when the metal ions bind to undesired sites, competing with other beneficial metal ions, or when adverse reactivity occurs due to their presence beyond the optimal concentration range. Conversely, concentrations below the optimum range lead to deficiencies, characterized by reduced physiological responses that can be fatal in severe cases. It is worth noting that Bertrand's plot specifically applies to essential trace elements and may not hold true for bulk elements or metal ions lacking known biological functions. Table 1.4 provides a compilation of diseases associated with deficiencies and excesses of various essential trace elements. While the adverse effects of deficiency and excess of essential metal ions are discussed in the respective chapters, this section focuses on toxicity caused by non-essential metals such as mercury, cadmium, lead, arsenic, and chromium. However, prior to delving into the discussion on toxicity, it is important to establish a clear definition of toxicity itself. Toxicity can be defined as the capacity of a chemical or substance to induce harmful effects when an organism is exposed to it, whether through accidental exposure or intentional administration [28].

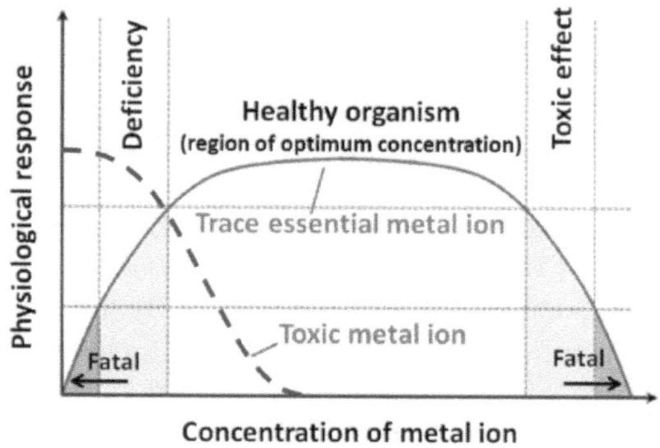

FIGURE 1.12 Bertrend's plot showing the dose-response relationship for essential metal ions and toxic metal ions. (Copyright permission from Elsevier, Ref. [3].)

TABLE 1.4
Abnormal Condition Due to Deficiency and Excess of Trace Essential Elements

Essential Elements	Diseases Due to Deficiency	Diseases Due to Excess
Chromium	Abnormal glucose metabolism	Carcinogenic
Manganese	Skeletal deformities	Ataxia
Iron	Anaemia	Siderosis
Cobalt	Pernicious anaemia	Coronary failure
Copper	Anaemia	Wilson's disease
Zinc	Dwarfism	Metal fume fever
Molybdenum	Cerebral atrophy	Carcinogenic

TOXIC EFFECTS OF MERCURY

Mercury, in the form of Hg(II) ions, is released into the environment through the weathering of red cinnabar (HgS), as well as the use of agricultural organomercurials (RHgX), which contribute to toxic waste. The binding of both RHgX and HgX_2 compounds to sulfhydryl groups in proteins can lead to alterations in protein structure and severe health issues, such as neurological diseases and kidney failure. However, the primary target organ of mercury toxicity is the brain. The devastating effects of mercury toxicity were exemplified by the Minamata disease outbreak in Japan in 1956, causing neurological symptoms like ataxia, numbness, muscle weakness, visual impairment, and damage to hearing and speech [5]. Methylmercury (MeHg), the predominant form responsible for the Minamata tragedy, is derived from the methylation of inorganic mercury in aquatic sediments and soils. MeHg is readily absorbed from the diet and distributes throughout the body within a few days, particularly accumulating in the brain by crossing the blood-brain barrier as Me-Hg-cysteine complex. It is believed that the thiol groups of SH-dependent enzymes are the primary targets of mercury. Surprisingly, selenocysteine exhibits a higher affinity for mercury than cysteine, impacting selenoenzymes crucial for protecting the brain and neuroendocrine system against oxidative damage [29]. Me-Hg can irreversibly inhibit these enzymes. The breakdown of selenoproteins provides selenide, which binds to mercury, forming HgSe and accumulating in cellular lysosomes. Excessive mercury presence leads to the formation of insoluble Hg selenides, preventing the bioavailability of selenium for essential physiological functions [5].

Mercury poisoning can be treated with decorporating agents like BAL (2,3-dimercaptopropanol) and N-acetylpenicillamine. Remarkably, bacteria resistant to mercury have developed a natural detoxification system [30]. Several gene products, including MerT and MerP, facilitate the specific uptake of mercury compounds and work alongside other proteins in the Mer operon to efficiently remove mercury from bacterial cells. Another enzyme, organomercury lyase encoded by the merB gene, plays a significant role in breaking Hg-C bonds. Additionally, mercuric reductase (MerA) features both redox-active and redox-inactive cysteines, enabling the selective reduction of Hg(II). MerA not only detoxifies mercury from the environment but also assists in clearing Hg^{2+} generated by the MerB protein from RHgX compounds [31]. Nature has devised an extraordinary system within these mercury-resistant bacteria to effectively detoxify mercury, highlighting the fascinating adaptability of these organisms.

TOXIC EFFECTS OF CADMIUM

Cadmium, being a soft Lewis acid, has shown a preference for ligands with soft donor atoms, such as sulfur, which allows it to displace zinc from proteins where the zinc is in a sulfur-rich environment. Due to the similarity in ionic radius/ionic potential of Ca^{2+}, Cd^{2+} can displace Ca^{2+} from calcium-binding proteins [32]. Furthermore, biomolecules involved in handling alkaline earth and transition metal ions, such as Mg(2+), Ca(2+), Zn(2+), Cu(2+/+), and Fe(3+/2+), are particularly sensitive to the presence of Cd(2+) due to their cationic sites, which can serve as binding sites for this toxic metal. These factors contribute to the toxicity mechanism associated with cadmium exposure. Cadmium toxicity has the potential to disrupt iron homeostasis, leading to imbalances in iron levels within the body [33]. Cd competes with iron for binding sites on proteins involved in iron transport, thereby inhibiting iron absorption, which leads to abnormal iron metabolism. Exposure to cadmium can lead to respiratory disease, emphysema, renal failure, bone disorders, and immunosuppression [34]. Cadmium is also known to promote oxidative stress and chemical modification of DNA or histones, such as DNA methylation, that changes the structure of chromatin without altering the DNA sequence. However, it severely affects the genetic information transfer process [35].

The devastating effect of cadmium toxicity was exemplified by a severe health crisis known as 'Itai-Itai' disease, which translates to 'it hurts-it hurts', among inhabitants of the Jinzu River basin in Toyama, Japan [36]. The prime cause was long-term exposure to high levels of cadmium in contaminated water and food. The disease is characterized by excruciating bone pain, skeletal deformities, kidney damage, and other debilitating symptoms. Cadmium toxicity disrupts calcium metabolism, leading to weakened bones and damaged kidneys.

Chelation therapy is the most commonly employed therapeutic strategy for cadmium toxicity. However, it also has a very narrow treatment window. In cases of acute toxicity, chelators such as $CaNa_2EDTA$ (EDTA: ethylenediaminetetraacetate) and meso-2,3-dimercaptosuccinic acid (DMSA) have been identified as having protective effects against cadmium toxicity [37]. On the other hand, no effective chelator has been identified for chelation therapy in cases of chronic cadmium toxicity.

TOXIC EFFECTS OF LEAD

Lead (Pb) is a non-essential element that can be toxic to the body, even in small amounts. Its toxicity poses a serious concern as it can have detrimental effects on organism health. The primary targets of Pb are proteins that require calcium (Ca) and zinc (Zn) for their proper function [38]. These include synaptotagmin, which acts as a calcium sensor in neurotransmission, and δ-aminolaevulinate synthase (ALAD), the second enzyme in the heme biosynthetic pathway, which requires zinc for its activation. Despite the notable size difference, Pb is capable of substituting Ca in synaptotagmin and Zn in ALAD [5]. Pb^{2+} and Zn^{2+} appear to compete for a single metal-binding site. Exposure to Pb disrupts neurotransmission by interfering with Ca^{2+}-binding proteins, impairing the detection of crucial Ca^{2+} signals necessary for neurotransmission. Furthermore, Pb^{2+} has been found to inhibit voltage-gated Ca^{2+} channels, hindering the increase in internal Ca^{2+} required for rapid Ca^{2+}-dependent vesicular release, thus interfering with neurotransmission [39]. Through these mechanisms, Pb negatively impacts various organ systems, including the nervous, hematopoietic, renal, endocrine, and skeletal systems. Lead also disrupts normal DNA transcription processes and causes bone disabilities. In infants and young children, Pb exposure can lead to behavioral and cognitive deficits during brain development, and there is mounting evidence that early-life Pb exposure may contribute to neurodegeneration later in life [5].

Pb poisoning is generally treated using $CaNa_2EDTA$, which chelates lead with high affinity and is subsequently excreted in urine [40]. Additionally, blood Pb levels can be significantly reduced by administering succimer and dimercaprol [41].

TOXIC EFFECTS OF ARSENIC

Arsenic (As), recognized as the ‘king of poisons’ and the ‘poison of kings’ since ancient times, is a highly toxic element with significant implications for human health and the environment [42]. Regulatory agencies, such as the International

Agency for Research on Cancer (IARC) and the European Union under Directive 67/548/EEC have classified arsenic and its compounds as mutagenic, carcinogenic, and harmful to the environment [41]. Arsenic exhibits various chemical forms, including metalloid (As0), inorganic species (As^{3+} and As^{5+}), organic compounds, and arsine (AsH_3). The toxicity of arsenic compounds follows a distinct order: organic arsenicals < As0 < inorganic species ($As^{5+} < As^{3+}$) < arsine, indicating varying degrees of toxicity [43]. Acute and chronic arsenic toxicity adversely affect the function of approximately 200 enzymes, notably those involved in cellular energy pathways, DNA replication and repair, and high-energy compound synthesis, including ATP. Similar to other heavy metals, arsenic can inhibit enzymes containing sulfhydryl groups, impairing their normal function. Furthermore, arsenic binds to the lipoic acid moiety of pyruvate dehydrogenase, leading to its inactivation. This inactivation can block the Krebs cycle and oxidative phosphorylation, resulting in a significant decrease in ATP production and causing cellular damage [44]. Additionally, arsenic-induced damage to the capillary endothelium increases vascular permeability, resulting in vasodilation and circulatory collapse. Such vascular impairment contributes to the detrimental effects of arsenic on various organ systems, including the nervous, hematopoietic, renal, endocrine, and skeletal systems [43].

TOXIC EFFECTS OF THALLIUM

Thallium (Tl) is a toxic heavy metal that was serendipitously discovered by Sir William Crookes in 1861 during an experiment involving the burning of dust from a sulfuric acid industrial plant. During the course of the experiment, Crookes observed a mesmerizing bright green spectral band that rapidly vanished. Intrigued by this phenomenon, he named the newly identified element 'Thallium', derived from the Greek word 'thallos', meaning young shoot or twig [45]. Thallium is classified as an electropositive element and exhibits two primary oxidation states: monovalent thallous (Tl+) and trivalent thallic (Tl3+) compounds. In its monovalent state, thallium has a propensity to form stable complexes with soft ligand donors such as sulfur and selenium.

Thallium ion (Tl+) has caused several occupational and accidental intoxications. Even trace amounts of Tl+ in the body are considered case of severe abnormality and demand immediate attention. A key factor contributing to thallium's toxic mechanism lies in the close chemical resemblance to Tl+ with potassium (K+) due to their closely matched ionic charges and radii. Such close similarity between Tl+ and K+ makes the cell membrane unable to distinguish between them. This lack of discrimination allows thallium to navigate cell membranes and interfere with K-dependent enzymes, including those involving the Na:K ATPase, a fundamental enzyme involved in maintaining cellular ion balance. As a consequence, thallous can mimic potassium in its movement within cells and accumulate intracellularly in mammals [45]. Furthermore, thallous exhibits a high affinity for sulfhydryl groups found in proteins and other biomolecules, further disrupting their normal functions.

Thallous intoxication can occur through multiple routes, such as inhalation, ingestion via contaminated food or hands, or skin contact. Due to its high toxicity and the fact that thallous salts are colorless, odorless, and tasteless, they have historically been used in homicidal poisoning and termed the 'poisonous poison'. Studies have indicated that thallous salts significantly affect cell metabolism, cause redox alterations, induce mitochondrial and kidney dysfunction, and activate apoptotic signaling pathways [46].

CONCLUSION

In conclusion, this chapter sheds light on the fundamental significance of metal ions in biological systems. Metal ions play pivotal roles in enzymatic reactions, cellular signaling, structural integrity, and gene expression regulation, which collectively drive various physiological processes. The classification of metal ions into essential, non-essential, and therapeutically relevant categories underscores their diverse roles in cellular function. The criteria for identifying metal ions as essential factors are thoroughly examined, encompassing factors such as bioavailability, kinetic and thermodynamic parameters, and their ability to fulfill specific biological roles. Furthermore, the chapter delves into the intricate interactions between metal ions and a variety of biological ligands, ranging from proteins and peptides to nucleic acids and small molecules. Additionally, the physiological response to the concentration of metal ions is explored, emphasizing the delicate balance required for proper cellular function. The chapter also highlights the toxic effects of non-essential metal ions, which are due to their interference with essential metal ions. Understanding the multifaceted roles of metal ions in biology is essential for advancing our knowledge of human health and disease.

REFERENCES

1. R. Crichton, Chapter 19 – Metals in medicine and metals as drugs. In *Biological Inorganic Chemistry* (Third Edition), Elsevier, Amsterdam, 2019, Pages 599–624, ISBN 9780128117415
2. R. Crichton, Chapter 1 – An overview of metals in biology. In *Biological Inorganic Chemistry* (Third Edition), Elsevier, Amsterdam, 2019, Pages 81–118, ISBN 9780128117415.
3. M. A. Zoroddu, J. Aaseth, G. Crisponi, S. Medici, M. Peana, V. M. Nurchi, The essential metals for humans: a brief overview, *J. Inorg. Biochem.* 2019, 195, 120–129.
4. W. Maret, The metals in the biological periodic system of the elements: concepts and conjectures, *Int. J. Mol. Sci.* 2016, 17, 66.
5. R. R. Crichton, Chapter 1 – Metal toxicity – An introduction. In R.R. Crichton, R.J. Ward, R.C. Hider (Eds.), *Metal Chelation in Medicine*, The Royal Society of Chemistry, Cambridge, 2016, Pages 1–23.
6. E. I. Ochiai, Principles in bioinorganic chemistry: Basic inorganic exercises, *J. Chem. Edu.* 1978, 55, 631.
7. S. J. Lippard, J. M. Berg, Chapter 5 – Choice, uptake and assembly of metal containing units in biology. In S. J. Lippard (Eds.), *Principles of Bioinorganic Chemistry,* University Science Books, 1994.

8. F. De Rienzo, R.R. Gabdoulline, M. C. Menziani, R.C. Wade, Blue copper proteins: A comparative analysis of their molecular interaction properties, *Protein Sci.* 2000, 9, 1439–1454.
9. E. I. Ochai, The interpretation of the EPR spectra of and the mechanism of B12-dependent reactions, *J. Inorg. Nucl. Chem.* 1975, 37, 351.
10. R. J. Abergel, J. A. Warner, D. K. Sukh, K. N. Raymond, Enterobactin protonation and iron release: structural characterization of the salicylate coordination shift in ferric enterobactin1, *J. Am. Chem. Soc.* 2006, 128, 8920–8931.
11. K. N. Raymond, E. A. Dertz, S. S. Kim, Enterobactin: an archetype for microbial iron transport, *Proc. Natl. Acad. Sci. USA* 2003, 100, 3584–3588.
12. https://www.open.edu/openlearn/mod/oucontent/view.php?printable=1&id=2497.
13. R. Crichton, Chapter 4 – Biological ligands for metal ions. In *Biological Inorganic Chemistry* (Third Edition), Elsevier, Amsterdam, The Netherlands, 2019, Pages 81–118, ISBN 9780128117415.
14. S. J. Lippard, J. M. Berg, Chapter 3 – Properties of biological molecules. In S. J. Lippard (Eds.), *Principles of Bioinorganic Chemistry,* University Science Books, Mill Valley, CA, 1994.
15. C. Calmettes, J. Alcantara, R.-H. Yu, A. B. Schryvers, T. F. Moraes, The structural basis of transferrin sequestration by transferrin-binding protein B, *Nat. Struct. Mol. Biol.* 2012, 19, 358–360.
16. N. Cox, A. Nalepa, M.-E. Pandelia, W. Lubitz, A. Savitsky, Pulse double-resonance EPR techniques for the study of metallobiomolecules, *Methods Enzymol.*, 2015, 563, 211–249.
17. R. Mandal, G. Dutta, From photosynthesis to biosensing: chlorophyll proves to be a versatile molecule, *Sensors Int.* 2020, 1, 100058.
18. S. A. Kerns, A. Biswas, N. M. Minnetian, A. S. Borovik, Artificial metalloproteins: at the Interface between biology and chemistry, *JACS Au* 2022, 2, 1252–1265.
19. R. H. Holm, P. Kennepohl, E. I. Solomon, Structural and functional aspects of metal sites in biology, *Chem. Rev.* 1996, 96, 2239–2314.
20. G. B. Jameson, J. A. Ibers, Chapter 4 – Biological and synthetic dioxygen carriers. In I. Bertini, H.B. Gray, S.J. Lippard, J.S. Valentine (Eds.), *Bioinorganic Chemistry*, University Science Books, Mill Valley, CA, 1994.
21. H. B. Gray, J. R. Winkler, Electron flow through metalloproteins, *Biochim. Biophys. Acta – Bioenerg.* 2010, 1797, 1563–1572.
22. M. Cassandri, A. Smirnov, F. Novelli, C. Pitolli, M. Agostini, M. Malewicz, G. Melino, G. Raschella, Zinc-finger proteins in health and disease, *Cell Death Discov.* 2017, 3, 17071.
23. E. -I. Ochiai, Why calcium? Principles and applications in bioorganic chemistry-IV, *J. Chem. Edu.* 1991, 68, 10.
24. S. J. Lippard, J. M. Berg, Chapter 11 – Atom and group transfer chemistry. In S. J. Lippard (Eds.), *Principles of Bioinorganic Chemistry*, University Science Books, Mill Valley, CA, 1994.
25. S. J. Lippard, J. M. Berg, Chapter 1 – Overview of bioinorganic chemistry. In S. J. Lippard (Eds.), *Principles of Bioinorganic Chemistry*, University Science Books, Mill Valley, CA, 1994.
26. R. Banerjee, Radical carbon skeleton rearrangements: catalysis by coenzyme B12-dependent mutases, *Chem. Rev.* 2003, 103, 2083–2094.
27. P. K. Bhattacharya, P. B. Samnani, *Metal Ions in Biochemical Systems* (Second Edition), CRC Press, Boca Raton, 2020.
28. K. S. Egorova, V. P. Ananikov, Toxicity of metal compounds: knowledge and myths, *Organometallics* 2017, 36, 4071–4090.

29. N. V. C. Ralston, L. J. Raymond, Dietary selenium's protective effects against methylmercury toxicity, *Toxicology* 2010, 278, 112.
30. S. J. Lippard, *Chapter 9 – Metal in medicine.* In I. Bertini, H.B. Gray, S.J. Lippard, J.S. Valentine (Eds.), *Brinorganic Chemistry*, University Science Books, Mill Valley, CA, 1994.
31. T. Barkay, S. M. Miller, A. O. Summers, Bacterial mercury resistance from atoms to ecosystems, *FEMS Microbiol. Rev.* 2003, 27, 355–384.
32. J.-M. Moulis, Cellular mechanisms of cadmium toxicity related to the homeostasis of essential metals, *Biometals* 2010, 23, 877.
33. Y. Fujiwara, J.-Y. Lee, H. Banno, S. Imai, M. Tokumoto, T. Hasegawa, Y. Seko, H. Nagase, M. Saoh, Cadmium induces iron deficiency anemia through the suppression of iron transport in the duodenum, *Toxicol Lett.* 2020, 332, 130–139.
34. L. Järup, A. Akesson, Current status of cadmium as an environmental health problem, *Toxicol. Appl. Pharmacol.* 2009, 238, 201.
35. B. Wang, Y. Li, C. Shao, Y. Tan, L. Cai, Cadmium and its epigenetic effects, *Curr. Med. Chem.* 2012, 26, 11–20.
36. M. M. Mehdikhanmahaleh, O. T.-Malazy, *Itai Itai Disease, Reference Module in Biomedical Sciences*, Elsevier, Amsterdam, The Netherlands, 2023.
37. S. W. Smith, The role of chelation in the treatment of other metal poisonings, *J. Med. Toxicol.* 2013, 9, 355.
38. H. A. Godwin, The biological chemistry of lead, *Curr. Opin. Chem. Biol.* 2001, 5, 223.
39. C. Xiao, Y. Gu, C. Zhou, L. Wang, M. Zhang, D. Ruan, Lead impairs GABAergic synaptic transmission in rat hippocampal slices: a possible involvement of presynaptic calcium channels, *Brain Res.* 2006, 1088, 93–100.
40. A. L. Wani, A. Ara, J. A. Usmani, Lead toxicity: a review, *Interdiscip. Toxicol.* 2015, 8, 55–64.
41. S. K. Park, M. S. O'Neil, P. S. Vokonas, D. Sparrow, R. O. Wright, B. Coull, H. Nie, H. Hu, J. Schwartz, Air pollution and heart rate variability effect modification by chronic lead exposure, *Epidemology* 2008, 19, 111–120.
42. D. K. Gupta, S. Tiwari, B. Razafindrabe, S. Chatterjee, Arsenic contamination from historical aspects to the present. In D. Gupta, S. Chatterjee (Eds.), *Arsenic Contamination in the Environment*, Springer, Berlin, Germany, 2017, Pages 1–12.
43. M. B. Mood, K. Naseri, Z. Tahergorabi, M. R. Khazdair, M. Sadeghi, Toxic mechanisms of five heavy metals: mercury, lead, chromium, cadmium, and arsenic, *Front. Pharmacol.* 2021, 12, 643972.
44. S. Shem, X. F. Li, W. R. Cullen, M. Weinfeld, X. C. Le, Arsenic binding to proteins, *Chem. Rev.* 2013, 113, 7769–7792.
45. S. G. Arzate, A. Santamaria, Thallium toxicity, *Toxicol. Lett.* 1998, 99, 1–13.
46. L. S. Chapul, A. Santamaria, M. Aschner, T. Ke, A. A. Tinkov, I. Tunez, L. O. Rico, S. G. Arzate, E. R. Lopez, Thallium-induced DNA damage, genetic, and epigenetic alterations, *Front. Genet.* 2023, 14, 1168713.

2 Basic Principles of Coordination Chemistry

INTRODUCTION

Bioinorganic chemistry is primarily based on the involvement of metal ions in biological systems. From an introductory perspective, it is important to know the basics of coordination chemistry to understand the concepts involved in the development of bioinorganic topics. The properties related to bond formation as per the hard and soft acids and base principle, the stability of enzymes, the favorable geometry of metal centers, and their preferable ionic states are essential. Other important features involve the discussion of electron transfer reactions, their functioning, the magnetic properties of complexes, and the distortions involved in various geometries. From this prospect, the stability of complexes and other details are discussed as follows:

HARD-SOFT ACID-BASE CONCEPTS

Metal ions present in our biological system exhibit preferential bonding toward ligands dictated by HSAB theory. This concept explains the preferential stability of complexes formed due to the combination of a metal (Lewis acid) and a ligand (Lewis base). As per the principle, hard cations form a stable compound with hard anion and vice versa. 'Hard' species are classified as small and less polarizable moieties whereas 'soft' corresponds to species that are larger in size and more polarizable [1,2]. A table is provided for hard acid-base preferences (Tables 2.1 and 2.2).

Metal ions are mostly embedded in biological systems; however, the ligands are primarily present in protein side chains, organic cofactors, and water. Several examples can be observed in our system to explain this concept effectively. For example, in the case of glutamate and aspartate amino acid residues, the carboxylate group prefers to coordinate with alkali and alkaline earth metals through carboxylate oxygen. Another suitable example to understand this notion is the bonding of the Cu(II) ion with the imidazole nitrogen of the histidine amino acid residue [3]. Metallothionein is another illustrative example that can be taken into account to understand this theory. The defined function of metallothionein is to protect the cell against the toxic effects of heavy metals. Nearly one-third of the amino acids in this class of proteins have cysteine residues with sulfhydryl groups, which can bind avidly with toxic metal ions such as Cd^{2+}, Hg^{2+}, and Pb^{2+}.

DOI: 10.1201/9781003432333-2

TABLE 2.1
Classification of Important Acid and Base as per HSAB Principle [1,2]

Metal (Lewis Acid)		
Hard Acid	**Border Line**	**Soft Acid**
H^+ Be^{2+} Ti^{4+}	Fe^{2+} Rh^{3+}	Cu^+ Cd^{2+} Pt^{2+}
Li^+ Mg^{2+} Cr^{3+}, Cr^{6+}	Co^{2+} Ir^{3+}	Ag^+ Hg^{2+} Pt^{4+}
Na^+ Ca^{2+} Mn^{2+}, Mn^{7+}	Ni^{2+} Ru^{3+}	Au^+ Hg_2^+ CH_3Mg^+,
K^+ Sr^{2+} Fe^{3+}, Co^{3+}	Cu^{2+} Os^{2+}	RSe^+, M^0 (metal
	Zn^{2+} Pb^{2+}	atom)

Ligand (Lewis Base)		
Hard Base	**Border Line**	**Soft Base**
NH_3, RNH_2	N_2, N_3^-, NO_2^-	H^-, R^-, CO, RNC
H_2O, OH^-, O^{2-}, ROH, RO^-	Aniline, imidazole	CN^-, SCN^-, R_2S,
CH_3COO^-, NO_3^-,	Br-	RS^-, R_2Se, I^-
PO_4^{3-}, CO_3^{2-}		

TABLE 2.2
Correlation between the Coordination Ability and Function of Biologically Relevant Metal ions [4,5]

Metal Ion	Coordination Ability	Movement	Function
V, Mn, Fe, Co, Ni, Cu, Mo	Strong	Low	Redox behavior
Mg^{2+}, Ca^{2+}, Zn^{2+}	Intermediate	Intermediate	Structural role
Na^+, K^+	Weak	High	Charge carrier

CHELATE EFFECT

Chelation is an important process to enhance the stability of coordination compounds. It refers to the coordination of two or more donor atoms present in a ligand, which results in the formation of a ring with unusual stability. Subsequently, the resulting complex encompasses at least one 'metallacycle' ring structure, which freezes the torsional mobility of the system and contributes to enhanced selectivity toward central metal ion. Due to the preference for sp^2- and sp^3-hybrid atoms at 120° and 109° bond angles, respectively, the optimum ring size of the metallacycles in chelate complexes ranges from four- to six-membered rings, depending on the size of the metal ions, whether small or very large. This stability arises due to favorable entropy changes. The importance of translational entropy is purely statistical and probabilistic in nature. To develop a better understanding, a simple example can be taken into account:

$$\left[Ni(NH_3)_6\right]^{2+} + 3\,en \Leftrightarrow \left[Ni(en)_3\right]^{2+} + 6\,NH_3 \tag{2.1}$$

The change in entropy would apparently be proportional to the difference between the number of molecules present when the reaction initiates and after the completion of the reaction. For example, in Equation 2.1, it is demonstrated that $[Ni(NH_3)_6]^{2+}$ reacts with ethylene diamine to form the compound $[Ni(en)_3]^{2+}$ (Equation 2.1). The above reaction takes place in the forward direction with an increasing number of moles. This, in turn, favors the formation of the chelated product instead of the hexamine compound of the monodentate ligand ammonia. The substitution of ammonia molecules with the chelating ligand ethylenediamine increases the number of molecules in the solution, resulting in an increase in entropy as per $\Delta S = nR\ln 55.5 = 33.4$ J/mol/K, where n is the number of chelated rings available in the resulting product [1,3].

In the same sequence, another important prospect in bioinorganic chemistry is the presence of macrocyclic ligands, e.g., porphyrin, corrin, etc. These macromolecules contain four coplanar pyrrole rings with their nitrogen donor atoms pointing toward the central metal ion. The subsequent metalloporphyrin and corrin compounds are thermodynamically stable, and the space developed through the rings is efficient enough to accommodate a wide range of metal ions. Chlorophyll (Mg) and Vitamin B12 (Co) are other notable examples based on the same concept [3,4].

IONIC RADII

Another deciding factor that governs the extent of bonding between a transition metal ion and a ligand system is the ionic radius. The ionic radius across the period decreases with an increasing Z_{eff} value. Subsequently, the enhanced Lewis acidity drastically boosts the stability of the complex through strong metal-ligand bonds. To generalize these characteristics, an order of stability has been provided, i.e., Irving-Williams series. The sequence of the metal ions is as follows [2]:

$$Mn^{2+} < Fe^{2+} < Co^{2+} < Ni^{2+} < Cu^{2+} > Zn^{2+}$$

pK_a VALUE OF COORDINATED LIGANDS

The presence of a positive charge on the active site (central metal) of an enzyme assists in stabilizing anionic biological ligands, i.e., water, imidazole, alcohol, and the carboxylates of Asp and Glu. The coordination of a metal ion with the mentioned ligands can be epitomized by pK_a values. The initial steps of various biological processes, with respect to the presence of various metals and their corresponding pK_a values [3], have been enlisted in Table 2.3 to demonstrate the importance of this effect.

TABLE 2.3
Effect of pK_a Value on the Coordination of Various Metal Ion [3]

Reaction	Metal Ion	pK_a
$[M\text{-}H_2O]^{2+} \rightleftharpoons [M\text{-}OH]^{+}$ (H^+ / H^+)	Mn^{2+}	11.1
	Cu^{2+}	10.7
	Zn^{2+}	10.0
	None	14.0
$[M\text{-}NH_3]^{2+} \rightleftharpoons [M\text{-}NH_2]^{2+}$ (H^+ / H^+)	Co^{2+}	32.9
	Ni^{2+}	32.2
	Cu^{2+}	30.7
	None	35.0
$[M\text{-}AcOH]^{2+} \rightleftharpoons [M\text{-}OAC]^{+}$ (H^+ / H^+)	Mg^{2+}	4.2
	Ca^{2+}	4.2
	Ni^{2+}	4.0
	None	4.7
$[M\text{-}Im]^{2+} \rightleftharpoons [M\text{-}Im]^{+}$ (H^+ / H^+)	Co^{2+}	4.6
	Ni^{2+}	4.0
	Cu^{2+}	3.8
	None	7.0

Notifying fact is that a higher charge on the metal center decreases the pK_a values. Thus, a trivalent cation lowers the value compared to a bivalent cation. Deprotonation is a crucial step in several enzymatic catalytic reactions to understand their functioning. The availability of two or more metals, as in (SOD), at an active site results in a significant lowering of pK_a values [3].

SUBSTITUTION REACTIONS

Substitution reactions in coordination compounds are those in which an atom, ion, or molecule in a complex is displaced by another metal ion or ligand. Substitution reactions are of two types:

i. Nucleophilic substitution reactions, denoted by S_N
ii. Electrophilic substitution reactions, denoted by S_E

Moreover, the reagents are also classified broadly into two types: nucleophilic and electrophilic reagents. In coordination compounds, the ligands are bonded to the central metal ion by making use of their lone pair of electrons. In such cases, the metal ion can be regarded as electrophilic and the ligand as a nucleophilic reagent. In general, it can be inferred that if the ligand of one complex is substituted by another ligand in the metal complex, the type of reaction is called nucleophilic substitution. However, when the metal is substituted by another metal, the reaction is called electrophilic substitution.

These substitution reactions can be further explained with the help of two basic mechanisms, i.e., associative and dissociative, termed as S_N2 and S_N1 substitution reactions. Associative reactions involve the slow formation of an intermediate seven-coordinated complex by the attachment of an addendum ligand to an octahedral complex (Equation 2.2) [2]. The over-crowding of ligands results in the disintegration of the intermediate species to afford a new substituted octahedral complex. In this mechanism, the rate-determining step involves the concentration of both reactant species.

$$[ML_5X] + Y \rightarrow [ML_5XY] \rightarrow [ML_5Y] + X \tag{2.2}$$

In contrast to the associative mechanism in the case of the S_N1 substitution reaction, the complex slowly dissociates to a species with a coordination number of 5 and square pyramidal geometry, which then converts to a stable trigonal bipyramidal configuration. This slow displacement of the ligand makes the metal ion electron-deficient, showing high susceptibility to the addition of an incoming ligand to yield a new substituted complex (Equation 2.3). Since the slow rate-determining step is dependent on the concentration of only one reactant species, the overall reaction is unimolecular.

$$[ML_5X] \rightarrow [ML_5] + Y \rightarrow [ML_5Y] \tag{2.3}$$

At a coordination number of 5, the favorable geometry could be either square pyramidal or trigonal bipyramidal in nature. It is quite complicated to distinguish their nature in biological systems due to the interaction of the incoming ligand with other nearby groups to provide stability. In the case of macromolecules, many conformational changes are involved in this reaction [2].

Electron Transfer Reaction

Two more feasible ways of electron transfer among the complexes are through the inner-sphere and outer-sphere mechanisms. Inner coordination electron transfer reactions require a bridging ligand to connect two metal ions of different coordination spheres to facilitate the electron transfer process [1,2]. To understand the mechanism, a very popular example is widely used, where the hexa-aqua complex $[Cr(H_2O)_6]^{2+}$ is oxidized by the pentaamine cobalt(III) complex $[Co(NH_3)_5X]^{2+}$ to afford the species $[Cr(H_2O)_5X]^{2+}$. The isolation of this species supports the transfer of the bridging group X^- from Co^{III} to Cr^{II} through an activated complex (Scheme 2.1).

Other than the presence of a bridging ligand, the nature of the complex involved in the reaction also plays a crucial role. As mentioned above, the reactant $[Cr(H_2O)_6]^{2+}$ is labile, whereas $[Co(NH_3)_5Cl]^{2+}$ is inert in nature. Conclusively,

$$[Cr(H_2O)_6]^{2+} + [Co(NH_3)_5Cl]^{2+} \longrightarrow [(H_2O)_5Cr^{II}\text{-}Cl\text{-}Co^{III}\text{-}(NH_3)_5]^{4+}$$

$$\updownarrow \text{ Electron Transfer}$$

$$[Cr(H_2O)_5Cl]^{2+} + [Co(H_2O)_6]^{2+} + 5NH_4^{+} \xleftarrow{\text{Hydrolysis}} [(H_2O)_5Cr^{II}\text{-}Cl\text{-}Co^{III}\text{-}(NH_3)_5]^{4+}$$

SCHEME 2.1 Reaction of $[Cr(H_2O)_6]^{2+}$ and $[Co(NH_3)_5Cl]^{2+}$ through inner electron transfer reaction.

$$[Cr(H_2O)_6]^{2+} + [Co(NH_3)_5Cl]^{2+} \longrightarrow [Cr(H_2O)_5Cl]^{2+} + [Co(H_2O)_6]^{2+} + 5NH_4^{+}$$

$[Cr(H_2O)_6]^{2+}$	$[Co(NH_3)_5Cl]^{2+}$	$[Cr(H_2O)_5Cl]^{2+}$	$[Co(H_2O)_6]^{2+}$
$t_{2g}^3\ e_g^1$	t_{2g}^6	t_{2g}^3	$t_{2g}^5\ e_g^2$
Labile	Inert	Inert	Labile

$$[Cr(H_2O)_6]^{2+} + [Co(NH_3)_6]^{2+} \longrightarrow [Cr(H_2O)_6]^{3+} + [Co(H_2O)_6]^{2+} + 6NH_4^{+}$$

$[Cr(H_2O)_6]^{2+}$	$[Co(NH_3)_6]^{2+}$	$[Cr(H_2O)_6]^{3+}$	$[Co(H_2O)_6]^{2+}$
$t_{2g}^3\ e_g^1$	t_{2g}^6	t_{2g}	$t_{2g}^5\ e_g^2$

SCHEME 2.2 Reaction of $[Cr(H_2O)_6]^{2+}$ and $[Co(NH_3)_6]^{2+}$ through outer sphere mechanism.

$$[Co^*(NH_3)_6]^{2+} + [Co(NH_3)_6]^{3+} \longrightarrow [Co^*(NH_3)_6]^{3+} + [Co(NH_3)_6]^{2+}$$
$$[Ru^*(NH_3)_6]^{2+} + [Ru(NH_3)_6]^{3+} \longrightarrow [Ru^*(NH_3)_6]^{3+} + [Ru(NH_3)_6]^{2+}$$

SCHEME 2.3 Example of self-exchange reactions (M and M* are isotopes).

$$[V(H_2O)_6]^{2+} + [Ru(NH_3)_6]^{3+} \longrightarrow [V(H_2O)_6]^{3+} + [Ru(NH_3)_6]^{2+}$$

SCHEME 2.4 Example of cross reactions.

for the inner-sphere electron transfer reaction to occur, one of the complex needs to be labile to permit forming of a bridge. It is worth mentioning that, in the absence of the chloride group in the reactant, the same reaction occurs through an outer-sphere electron transfer mechanism (Scheme 2.2).

Outer-sphere electron transfer reactions take place among complexes that do not undergo substitution. The electron transfer in these redox chemical reactions occurs through the intact coordination sphere of metal ions. If the electron transfer takes place between complexes of the same metal ion, it is called a homonuclear (self-exchange; Scheme 2.3) electron transfer reaction. In contrast, electron transfer between complexes of different metal ions is referred to as a heteronuclear (cross-reactions) [2]. In these reactions, no new bonds are formed or broken (Scheme 2.4).

Marcus proposed that in the outer-sphere mechanism, the following steps are involved:

1. Reactants diffuse together to form an outer-sphere complex called the precursor complex, in which the coordination spheres of both reactants remain intact.
2. Activation of the precursor complex initiates by changing the bond distances around each metal to resemble product-like state.
3. Reorganization of the activated complex triggers the formation of the successor complex through electron transfer.
4. The process of diffusion results in the dissociation of the successor complex into the product.

The formation of the precursor complex and the dissociation of the successor complex are fast. Marcus suggested that electron transfer should be adiabatic in nature since electronic transitions are much rapid compared to nuclear motion. Therefore, very fast electron transfer takes place once the internuclear distance becomes appropriately adjusted in the successor complex. The complex that gets oxidized during the electron transfer usually adopts shorter metal-ligand bonds because of the higher oxidation state, whereas the reduced complex achieves shorter metal-ligand distances in the anticipation of the lower oxidation state.

ELECTRONIC AND GEOMETRIC STRUCTURE OF METAL IN BIOLOGICAL SYSTEMS

Metals from Group 1 to Group 12 adopt various geometries in biological systems depending on their coordination number. Group 1 and Group 2 metals more frequently adopt a coordination number of six with octahedral being the favorable geometry [5,6]. However, the presence of variable oxidation states in transition elements has made this area a greater focus for researchers. Distorted geometries with respect to bond lengths and bond angles have been commonly reported in biological systems. The geometries of Group 1 and Group 2 metals, along with their respective oxidation states and functions, are tabulated in Table 2.4.

Simultaneously, the role of transition metals is well established in biological systems. Except for the first two metals, scandium (Sc) and titanium (Ti), of the first transition series, the occurrence of all other elements with drastic structural variety in thousands of metalloproteins has been widely reported. The geometric orientations of these metals define enzyme active sites to facilitate enzymatic reactivity as biological oxidation-reduction mediators. In the same context, molybdenum (Mo) is the only element of the second transition series that serves this function, whereas complexes of technetium (Tc), palladium (Pd), platinum (Pt), gold (Au), gadolinium (Gd), and other lanthanide metals have proven utilities for medical purposes. To gain a better understanding of the electronic nature of transition metals in various enzymes, we need to know about the valence d-electron of metal center. This information can be obtained by subtracting the oxidation state

TABLE 2.4
Existing Oxidation State and Function of Alkali and Alkaline Earth Metal in Biological Systems

Metal	Oxidation State	Coordination Number	Geometry	Function
Sodium	1^+	6	Octahedral	Charge carrier, Osmotic balance Nerve impulse
Potassium	1^+	6, 8	Octahedral	Charge carrier, Osmotic balance Nerve impulse
Magnesium	2^+	6	Octahedral	Structure in hydrolases, Isomerase Phosphate transfer
Calcium	2^+	6	Octahedral	Structure in hydrolases, Isomerase Phosphate transfer

of the metal and the atomic number of preceding noble gas from the atomic number of the metal center. The formulation of this concept is mentioned as follows:

Number of d electron = Atomic number of transition element – Oxidation of state of transition element – atomic number of preceding noble gas [7].

$$Mn(III) = 25 - 3 - 18 = 25 - 21 = 4$$

$$Cu(I) = 29 - 1 - 18 = 29 - 19 = 10$$

Partial filling of d electrons results in variable oxidation states, which in turn are responsible for mutable geometry with different coordination spheres. Nature and number are the main important features that decide the resulting geometry of the central metal. In this order, the notifying coordination number ranges from 3 to 6 with their adaptable geometry trigonal planner to octahedral (Figure 2.1). The wider flexibility has been reported in coordination numbers 3 and 4 in which the coordination number 3 displays trigonal planar, pyramidal, and 'T' shape configuration, whereas coordination number 4 shows square planar, tetrahedral, and see-saw geometry (Table 2.5).

As previously discussed, distortion does exist from the idealized structure. Although the available d orbitals on a free metal ion exhibit degenerate energy levels, in the presence of ligands, repulsion is observed, which results in the removal of their degeneracy depending on the ligand environment. The splitting of the d orbitals takes place with respect to the barycenter of the unsplit d electron energy levels (Figure 2.2). Basically, a metal ion placed at the center of a coordination polyhedron with a defined set of ligands alters the energy levels of the d orbitals from those found in the free metal ion. This conceptual prospect is

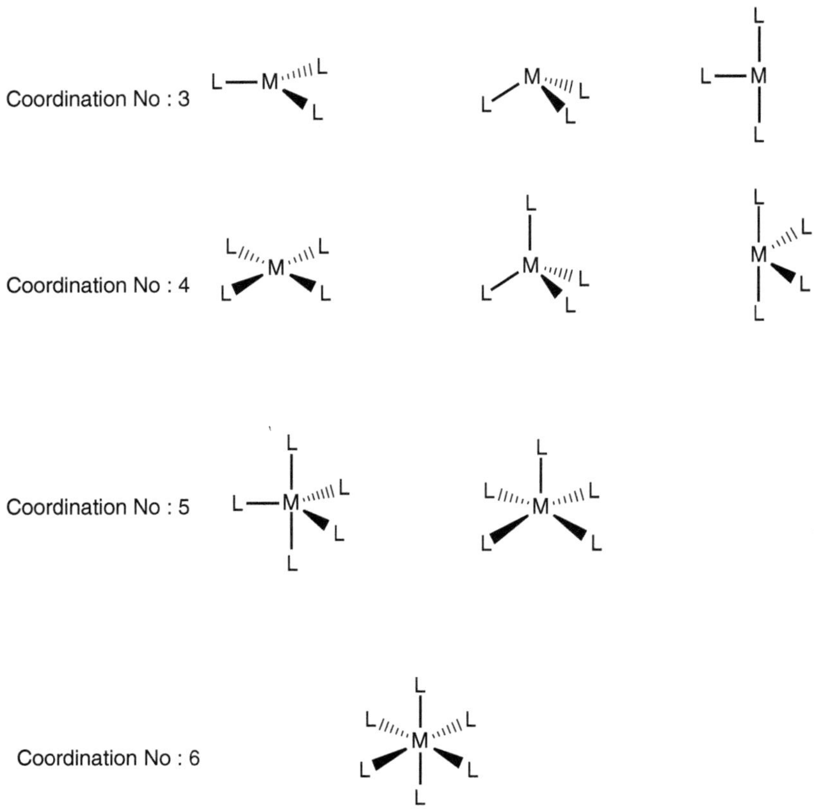

FIGURE 2.1 Coordination geometries from coordination number 3–6.

TABLE 2.5
Possible Geometry with Respect to their Coordination Number

Coordination Number	Geometry	Bond Angle
3	Trigonal planar	120°
	Pyramidal	90°< θ<109.5°
	T-shaped	90°, 180°
4	Square planar	90°
	Tetrahedral	109.5°
	See-Saw ("K") shaped	Ideal ax-ax 180°, eq-eq 120°
5	Trigonal pyramidal	90°, 120°
	Square pyramidal	<90°
6	Octahedral	90°

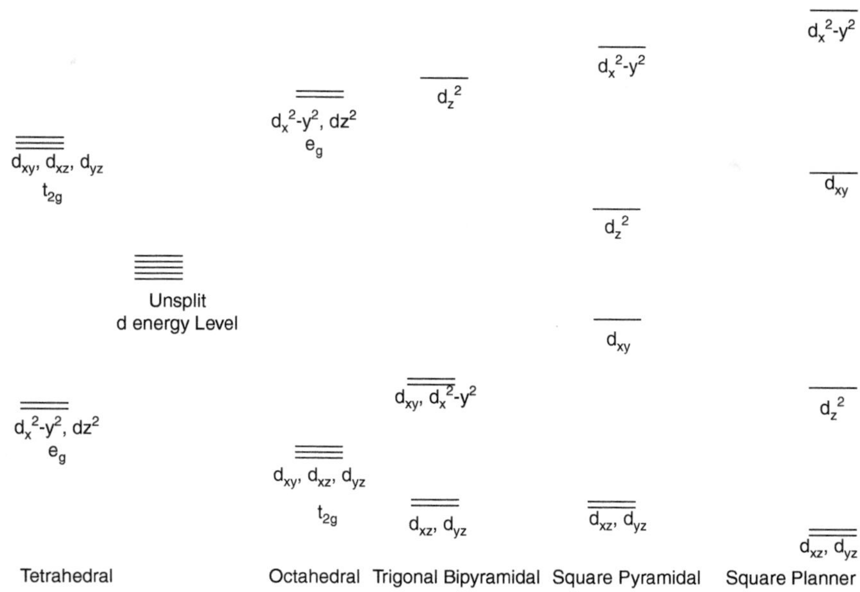

FIGURE 2.2 Splitting of degenerate d orbitals under various ligand fields.

called ligand field splitting, which can be best explained by the following energy level diagram as mentioned below [7].

The ligand field splitting diagram is very useful to understand the physical properties such as optical spectra, electron spin resonance and magnetism of the central metal ion available in the protein. To understand the splitting of degenerate d orbitals under the influence of approaching ligands, it is necessary to imagine the direction of the incoming electron density of the ligand sphere along the x, y, and z axes (Figure 2.3). Since, in the case of octahedral geometry, the ligands approach the metal system along the axis, the d orbitals oriented along the axis, i.e., $d_x^2 - y^2$, d_z^2, experience more repulsion compared to the other three oriented in between the axes (Figure 2.4). Consequently, the orbitals e_g ($d_x^2 - y^2$, d_z^2) are uplifted 6 Dq above the barycenter, and the remaining three, i.e., t_{2g} orbitals, are moved −4 Dq below average. The energy pattern gets exactly reversed in the case of a tetrahedral field, where the incoming ligands approach the metal center between the axes. The energy difference between these two sets of orbitals (t_{2g} and e_g) is greater in octahedral geometry compared to tetrahedral [2,6,7]. The two prominent reasons behind this nature are:

1. In the case of tetrahedral geometry, the number of ligands is four instead of six, as octahedral configuration. Altogether, the ligand field is only two-thirds the size; hence, the ligand field is also two-thirds in magnitude.

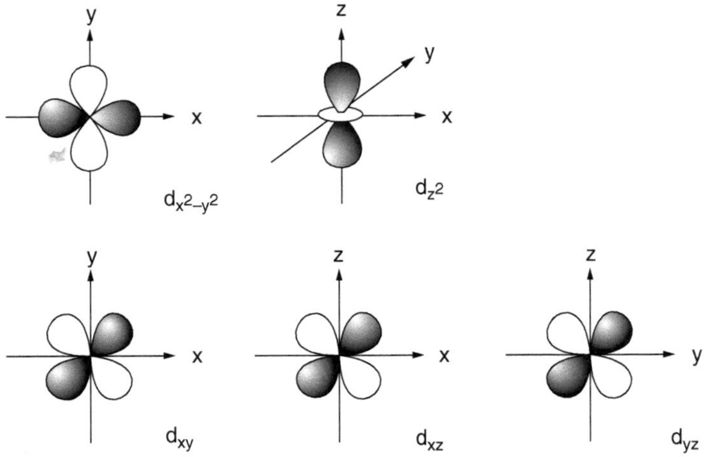

FIGURE 2.3 Shape of d orbitals.

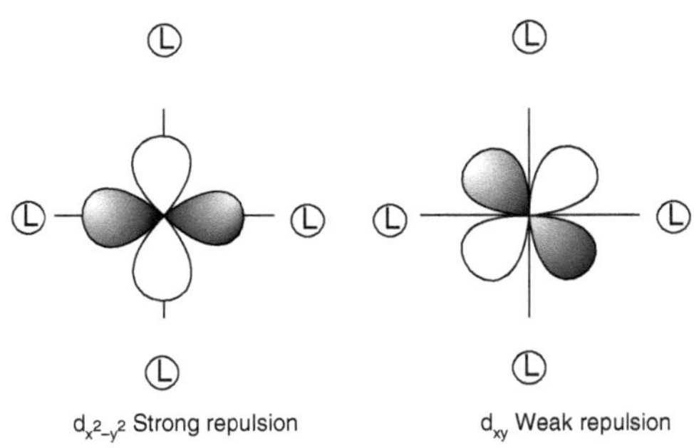

FIGURE 2.4 Representation of repulsion between d orbitals of metal and ligand in an octahedral environment.

2. Secondly, the direction of the ligands does not coincide with the orientation of the metal orbitals in tetrahedral geometry. This also reduces the crystal field splitting to two-thirds of its size.
3. Square planar complexes are another popular category for compounds with a coordination number four. The crystal field splitting diagram for these complexes is possible by the removal of *trans*-axial ligands from an octahedral complex to infinity. As the ligands along the *z*-axis are removed, the d orbitals with a *z*-component will be subjected to less electrostatic repulsion from the ligand. Hence, the energies of d_z^2, d_{xz}, and d_{yz} will be lowered in energy. There will be a corresponding increase in the energies of the other orbitals with respect to barycenter. Basically, in a square planar complex with ligands oriented in the *xy*-plane, i.e., the d_{x2-y2} orbitals will be the highest in energy, as the lobes point directly toward the ligands. The d_{xy} orbital, lying in the same plane, should be

the second highest in energy. The d_z^2 orbital appears next in energy, but in some cases, it may lie below the d_{xy} and d_{yz} orbitals. The crystal field splitting for square planar complexes has been noted to be higher than for octahedral complexes because the d_{xz} and d_{yz} orbitals interact with only two ligands in the square planner [2,6], while in the octahedral complex this interaction takes place with four ligands. The exact quantification can be correlated with the following equation:

$$\Delta_{sp} = 1.3\ \Delta_{oh}$$

Square planar geometries prominently occur in the d^8 transition metals of the 4d and 5d series, whereas they are less commonly reported in Ni(II) species with strong field ligands. A famous example of a square planar complex is *cis*-dichlo rodiammineplatinum(II), which is applied as an anticancer agent. The splitting of d orbitals contains the lower set of energy by 2/5 Δ_o as well as the higher set of energy by 3/5 Δ_o relative to the barycenter of the d orbitals. The octahedral complex can thus be said to stabilize to the extent of 0.4 Δ_o. The general formula to calculate the stabilization of various electronic configurations is shown in Figures 2.3 and 2.4.

$$\begin{aligned} \text{CFSE} = &\left(-0.4 \times \text{No of electron in } t_{2g} \text{ orbital}\right) \\ &+ \left(0.6 \times \text{No of electron in } e_g \text{ orbital}\right) + \text{pairing E} \end{aligned}$$

Let us take the example of a d^4 system of the octahedral environment, where two possibilities arise (Figure 2.5): either the electron may enter the higher energy e_g level, or it may pair with another electron in one of the t_{2g} orbitals [2,7]. The actual configuration adopted will, of course, be the one with the lowest energy and will depend on the relative magnitude of Δ_o and P:

$$\begin{aligned} \text{For weak field}\left(\text{High Spin}\right) &= t_{2g}{}^3 eg^1 \\ &= (-0.4 \times 3 + 0.6 \times 1) + 0 \times P \\ &= -0.6\ \Delta_0 \\ &= -0.6\ \text{Dq} \end{aligned}$$

$$\begin{aligned} \text{For strong field}\left(\text{Low Spin}\right) &= t_{2g}{}^4 eg^0 \\ &= (-0.4 \times 4 + 0.6 \times 0) + 1 \times P \\ &= -1.6\Delta_0 + P \\ &= -1.6\,\text{Dq} + P \end{aligned}$$

In the same context, the nature of ligands plays a decisive role in defining the magnitude of crystal field splitting of d orbitals. The effect of many ligands present in a complex has been studied, and the outcomes correspond that the variation

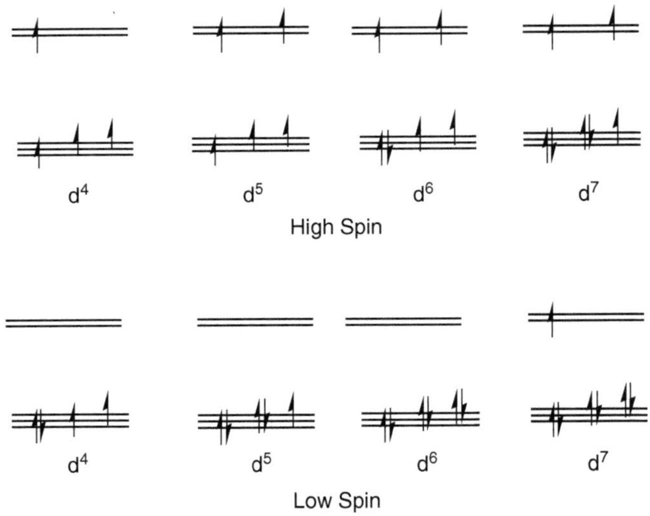

FIGURE 2.5 Filling of high-spin and low-spin d electron configurations in an octahedral field.

in the magnitude of the crystal field splitting with the nature of the ligand follows a regular order, known as the spectrochemical series [7]. The series is mentioned below, showing the order in which they produce an increasing energy gap:

$$I^- < Br^- < S^{2-} < Cl^- < NO_3^- < OH^- < RCOO^- < H_2O < RS^- < NH_3$$
$$< \text{imidazole} < \text{en} < \text{dipy} < \text{phen} < NO_3^- < \text{phosph} < CN^- < CO.$$

The probable explanation for the above series of ligands can be offered based on their polarizability, dipole moment, Π bonding properties, and electronegativity. Jorgensen proposed an approximate estimation of the value of Δ for an octahedral complex by considering it as the direct product of two independent factors:

$$\Delta = f \times g \times 10^{-3}\ \text{cm}^3$$

The quantity f describes the field strength of a ligand with respect to the water, which ranges from 0.7 (for bromide) to 1.7 (for strong field CN^-). The g factor is the characteristic of the metal that varies from 8,000 to 36,000 cm^{-1}. The field strength of the ligand also assists in predicting the geometry of the complex. For example, $[NiCl_4]^{2-}$ adopts the tetrahedral geometry with a small energy gap, whereas $[Ni(CN)_4]^{2-}$ occurs in a square planar shape due to large splitting. In octahedral and tetrahedral fields, the nature of the ligand is also helpful in determining the magnetic properties of the metal ion. In general, weak field ligands favor the high-spin orientation, whereas strong field ligands support low-spin configuration. That is the reason why, in general, the d^4 to d^7 electronic configurations of metal complexes in the first-row transition metal series exhibit paramagnetic behavior, with only d^6 complexes with strong field ligands opting for diamagnetic nature. Detection of paramagnetism (unpaired spin) and diamagnetism in bioinorganic field is very helpful in determining the coordination environment around the active sites of enzymes. For example, in the case of type III copper enzymes, two d^9 copper (II) centers couple anti-ferromagnetically, resulting in the loss of the expected paramagnetic behavior of the enzyme active center. Secondly, the d^6Fe(II) center in the case of hemoglobin sequences the movement of high-spin and low-spin configurations, affecting the placement of iron in or out of the plane of the porphyrin ring [3,8].

The Jahn-Teller effect explains why certain geometries undergo distortion. For example, six-coordinated octahedral complexes can show distorted octahedral or tetragonal geometries. This states that non-linear molecular systems possessing degenerate electronic states will be unstable and will undergo distortion to form a system of lower symmetry and lower energy, which removes the degeneracy. The distortion mainly arises due to unsymmetrical t_{2g} and e_g orbitals. One common example is complexes with a d^9 configuration, such as the aqueous solution of $[Cu(NH_3)_4]^{2+}$, in which trigonal distortion is so strong that a square planar complex results. The two possible arrangements for this configuration are [1,7,8]:

$$\text{I)}\ t_{2g}{}^{6}\left(d_z^2\right)^2\left(d_{x2-y2}\right)^1, \qquad \text{II)}\ t_{2g}{}^{6}\left(d_z^2\right)^1\left(d_{x2-y2}\right)^2$$

Since the t_{2g} orbitals are completely filled, distortion is caused by the incomplete filling of e_g orbitals. In configuration (i), d_z^2 orbitals are directed more toward the ligands, offering more shielding of the Cu^{2+} nucleus than half-filled d_{x2-y2} orbital projected on the xy-plane. Thus, the ligands on the x- and y-axes experience higher nuclear charge, while those on z axis experience lower effective nuclear charge. As a result, the ligands on x- and y-axes are drawn closer to the metal center, and those on the z-axis move outward. Thus, the ligand existing along z-axis would be at a greater distance, and the remaining four co-planar ligands would be at shorter distances. The exact opposite behavior can be observed in the second configuration (II).

CONCLUSIONS

The above-discussed text summarizes coordination chemistry, which paves the way for understanding bioinorganic chemistry. It is very important to understand the behavior of metals, which is greatly influenced by their surroundings. We shall also observe many distorted geometries of metal systems along with their coordination with ligands containing nitrogen and oxygen as donor atoms in various biological systems. These basics will definitely contribute to the development of mimics or models of biological system, intermingling the perspectives of inorganic and bioinorganic chemistry.

REFERENCES

1. J. E. Huheey, E. A. Keiter, R. L. Keiter, O. K. Medhi, *Inorganic Chemistry: Principles of Structure and Reactivity* (Fifth Edition), Pearson, India, 2006, Pages 1–886.
2. G. L. Miessler, G. A. Tarr, *Inorganic Chemistry* (Third Edition), Pearson, India, 2008, Pages 1–720.
3. S. J. Lippard, J. M. Berg, *Principles of Bioinorganic Chemistry*, University ScienceBooks, Mill Valley, CA, 1994.
4. J. A. Cowan, *Inorganic Biochemistry: An Introduction* (Second Edition), Wiley-VCH, New York, 1997.
5. R. M. Roat-Malone, *Bioinorganic Chemistry: A Short Course*, Ellis Horwood Limited, Halsted Press, New York, 2002, Pages 1–365.
6. R. W. Hay, *Bio-Inorganic Chemistry*, Ellis Horwood Limited, Halsted Press, New York, 1984.
7. W. Kaim, B. Schwederski, A. Klein, *Bioinorganic Chemistry: Inorganic Elements* in *the Chemistry of Life*, Wiley & Sons, Ltd, Hoboken, NJ, 2013, Pages 1–426.
8. J. Emsley, *Nature's Building Blocks*, Oxford University Press, Oxford, 2001.

3 Biochemistry of Alkali and Alkaline Earth Metals

INTRODUCTION

The higher abundance of alkali and alkaline earth metals in the Earth's crust and seawater has predestined these cations for noncatalytic functions. These cations commonly exist in the form of 'electrolyte' along with anions such as chloride. Apart from these four elements, the rest of the elements discussed in this book are characterized as 'trace metals'. However, these series do not belong to this category and are treated as major elements in the human body. Despite their chemical similarities, alkali and alkaline earth metals exhibit vast differences in biological functions. These metals signify their presence in elemental composition of the human body. The following functions are governed by these mono- and bivalent cations.

1. **Maintain Hydration and Osmotic pressure**

   The main function of these cations is to maintain the osmotic pressure of different body fluids to keep the body hydrated. Osmotic pressure completely depends on the total carbon present in the fluid. However, sodium serves as the chief cation and plays the dominant role in this process.
2. **Transport of CO_2**

   All these cations including chloride ions play a crucial role in the gaseous transport of CO_2.
3. **Neuromuscular role**

   Both sodium and potassium play a crucial role in maintaining normal neuromuscular irritability [2–4]. Their correlation can be expressed with the following equation:

$$\text{Irritability} \propto \frac{[\text{Na}^+]+[\text{K}^+]}{[\text{Ca}^{2+}]+[\text{Mg}^{2+}]+[\text{H}^+]}$$

   Ratios of various cations and the total absolute amount present in the body have significant importance.

DOI: 10.1201/9781003432333-3

4. **Storage of protein and glycogen**
   Potassium enters the intracellular fluid during cell growth and repair. The deposition of 1 g of cell protein requires the retention of 0.4 M equivalent of potassium. Similarly, 1 g of glycogen present in the liver and muscle needs about 0.15 M equivalent of potassium in the intracellular fluid. This is why insulin and glucose, when administered to diabetic patients to enhance glycogenesis, may lead to hypokalemia.
5. **Antibiotics**
   Alkali metal ions are weak Lewis acids and form weak complexes. The encapsulation of a cation by crown ethers is selective and depends on the cation's size. Some natural ion bearers of alkali metals have O-donor atoms from carboxylates and cyclic peptides. Some common antibiotics, such as nonactin and valinomycin, have formation constants for Na^+, K^+, and Cs^+ of 210, 2×10^4, and 400, respectively [2,4] (Figure 3.1).

Excretion

Potassium, sodium, and chlorine are excreted mainly by the kidneys in urine, to a lesser extent by the skin in the form of perspiration, and in small quantities in the digestive fluid by the gastrointestinal tract. Under normal conditions, the excretion of these electrolytes is equal to their intake. The most significant roles of these ions are discussed in details in Table 3.1.

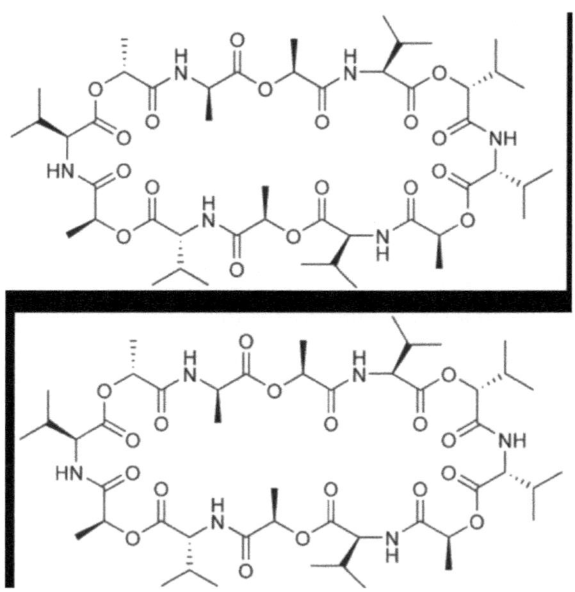

FIGURE 3.1 Structure of (a) valinomycin and (b) nonactin. (Reused with permission from Ref. [1]; Copyright (2023) Elsevier.)

TABLE 3.1
Concentration Difference and Resulting Electrochemical Intact

Concentration	Na^+	K^+	Ca^{2+}	Cl^-
Extracellular concentration (mmol/L)	150	5	1	120
Intracellular concentration (mmol/L)	15	150	10^{-5}	6
Equilibrium potential (mv)	+60	−90	+150	−80

SODIUM – POTASSIUM PUMP

Normally, cells are covered by a 70 Å thick double membrane that is separated by lipids. These membranes are impermeable to cations, and encapsulation is the only process by which cations can pass through the lipid layer [5,6]. Encapsulated cations present a lipid-soluble surface to the membrane.

The main function of the Na^+-K^+ pump came into light in 1957 by Danish scientist J. C. Skou. He discovered the Na-K-ATPase enzyme, with a molar mass of 110 kg/mol responsible for regulating the flow of Na^+/K^+ ions across the membrane [5–8]. In most animal cells, the concentration of K^+ and Na^+ is about 0.15 M and 0.01 M inside the cell, respectively, while the concentration of Na^+ and K^+ is about 0.15 M and less than 0.004 M outside the cell. To maintain these concentrations across the cell membrane, the Na^+/K^+ pump is required. The energy for this process is provided by the hydrolysis of ATP (Figure 3.2). Two-thirds of the energy spent on this transport is supplied by kidney and brain cells. In general, one ATP molecule is hydrolyzed in the transport of three Na^+ ions out of the cell and $2K^+$ into the cell.

MECHANISM

In the first step, ATP along with three sodium ions binds to the inside of the membrane, and the enzyme gets phosphorylated, inducing conformational changes called inversion. This process expels the Na^+ ions outside the cell membrane. The attachment of K^+ triggers dephosphorylation, and the loss of ATP induces conformational changes that bring two K^+ into the interior of the cell (Figure 3.3). This movement builds up a charge gradient across the cell membrane, with the outer surface becoming relatively positively charged [7,8].

The selective complexation of Na^+ and K^+ across the cell membrane is a unique feature from an inorganic perspective set. It is interesting that the same enzyme selectively binds to Na^+ first, then switches to K^+ with some conformational changes. Normally, Group-I cation complexes are not stable, so selectivity mainly depends on coulombic forces, which are directly related to ionic radii. Stronger bases than water complex with Na^+, while weaker bases than H_2O preferentially coordinate with K^+, as the displacement of water is more feasible for softer cations [7,8].

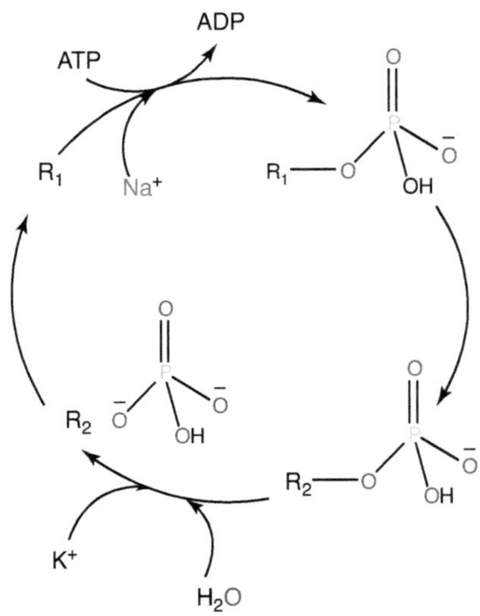

FIGURE 3.2 Sketch of sodium-potassium pump.

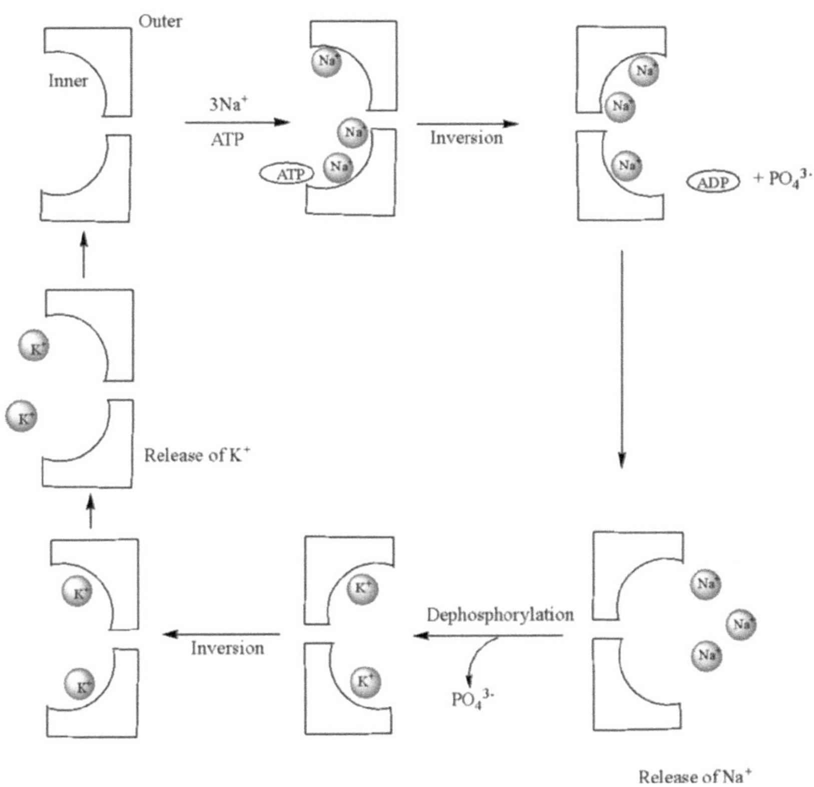

FIGURE 3.3 Steps involved in Na-K pump.

INHIBITOR

A trace amount of vanadate concentration inhibits the operation of ATPase. The difference between phosphate and vanadate ions is that the attack of water on phosphate results in a penta-coordinated phosphate moiety, which easily breaks down to complete hydrolysis. However, the penta-coordinated vanadate ion does not break down, which thwarts the conformational changes. As a consequence, K^+ remains outside the cell membrane [7].

CHLOROPHYLL

J. B. Caventou and P. J. Pelletier jointly isolated chlorophyll in 1817, and the presence of magnesium as the central core of chlorophyll was first established in 1906. Later, the expanded structure was explored by Hans Fischer as an active pigment in the photosynthesis process [4,9]. Chlorophyll appears green, is insoluble in water, but soluble in common organic solvents. The primary function of chlorophyll is to absorb energy strongly from light in the blue and red wavelengths of the electromagnetic spectrum. The maximum reflection of green light from the cell walls of plant leaves causes the green appearance of chlorophyll. Chemically, chlorophyll consists of two parts: the head, which is a tetrapyrrole porphyrin ring, and the tail, which is a long hydrocarbon chain of phytol attached to the porphyrin ring. Appearances in red, orange, yellow, brown, and green, etc., are related to chlorophyll. Carotenes are orange-red, and xanthophylls contain oxygen. The energy absorbed by carotenoids is transferred to chlorophyll *a*, which is utilized by photosynthesis [9].

STRUCTURE

The structure of chlorophyll consists of magnesium as a central atom embedded in the cavity of a porphyrin ring, similar to hemoglobin, but the basic difference between them is that hemoglobin contains an iron atom as its central atom (Figure 3.4). Chlorophyll is polar in nature, with the porphyrin ring forming the head and the protein chain forming the phytol tail, giving it a kite-like appearance [9]. The four nitrogen atoms of the porphyrin ring coordinate with magnesium to form a rhomboid shape. The head is embedded in the thylakoid membrane due to its hydrophilic nature, while the tail is lipophilic and wraps around the lipid layer of the thylakoid membrane. The diameter of the rhomboidal plane is about 15 Å, and the diameter of the phytol tail is approximately –20 Å. Depending on the presence of alkyl groups in the porphyrin ring, chlorophyll is classified into various categories, such as Chlorophyll a, b, c, d, etc. [4,9].

The absorption spectrum of chlorophyll shows a band in the red region and a second band in the blue to near-UV region, which are characteristic of the porphyrin ring. Both bands arise from the promotion of an electron from the porphyrin π HOMO to the π^* LUMO. The non-absorbing region between these two bands accounts for the appearance of green color.

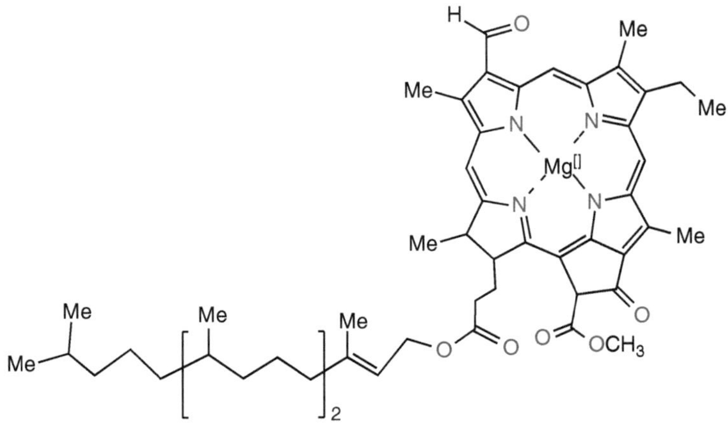

FIGURE 3.4 Structure of chlorophyll.

Chlorophyll-a has a yellow-green appearance, with the formula $C_{55}H_{70}O_6N_4Mg$, and is present in bryophytes, green algae, and vascular plants. During chemical reactions, chlorophyll-b absorbs light at different wavelengths and transfers it to chlorophyll-a through resonance. Chlorophyll-b, also called an accessory pigment, assists in the efficient absorption of light by broadening the spectrum of light absorbed. Accessory pigments behave like antennas, collecting light from various parts of the spectrum that chlorophyll alone cannot. The chain of reactions involved in the process initiates from the absorption of light and its conversion to chemical energy. Both the accessory pigments and the reaction center combine to form a photosystem. The basic difference between chlorophyll-a and chlorophyll-b is the replacement of a $-CH_3$ group with a $-CHO$ group in chlorophyll-b (Figure 3.5).

Photosynthesis

The day-to-day increasing awareness of the limited resources of fossil fuels and the carbon accumulation in environment caused by their burning has been a major motivation for understanding the functioning of photosynthesis. It is the fundamental process for the existence of life on Earth. The conversion of water and carbon dioxide into carbohydrates and oxygen using sunlight as an energy source is called photosynthesis. The formation of carbohydrates involves the reduction of CO_2 and the oxidation of H_2O to O_2. To complete this photochemical reaction, two photosystems are required, i.e., Photosystem I and II (PSI and II). This conversion occurs in the chloroplasts present in green leaves. It is worth noting that the primary step of the photosynthesis reaction is a type of outer-sphere mechanism.

$$nCO_2 + nH_2O \xleftarrow{hv} C_6H_{12}O_6 + 6O_2$$

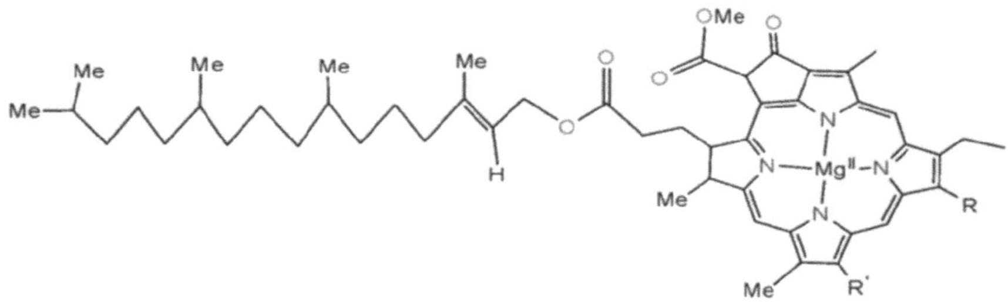

R	R'	Chlorophyll
CH_3	C_2H_5	Chlorophyll-a
CHO	C_2H_5	Chlorophyll-b
CH_3	CHO	Chlorophyll-c

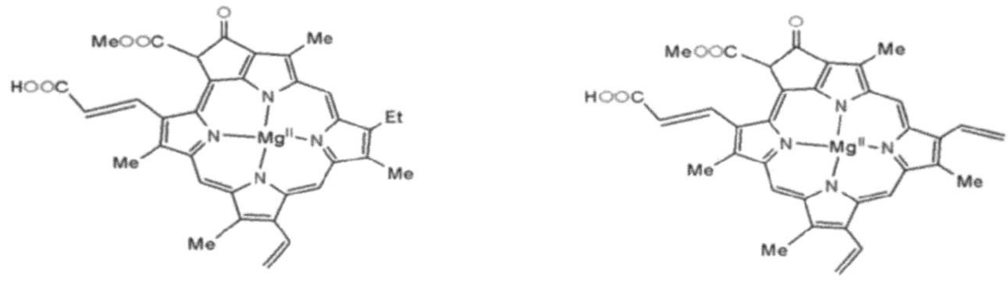

Chlorophyll-c_1 Chlorophyll-c_2

FIGURE 3.5 Structure of various chlorophyll.

SCHEMATIC PROSPECT

A set of cyclic events involved in the photosystem is demonstrated as follows (Figure 3.6):

1. Let us assume the resting state G_0, which is oxidized to G_1 by the loss of one electron along with a proton [9].
2. G_1 is further oxidized to G_2 with oxidation of one electron.
3. The third step, oxidation from G_2 TO G_3, involves the loss of electrons, which is accompanied by two moles of water producing oxygen and the expenditure of protons.
4. G_3–G_4 conversion takes place with the release of O_2, and the Mn enzyme is reduced by four equivalents of electrons. Due to the lack of a single crystal, the detailed structure of clusters involved in the oxidation process is not clear, but an oxo-bridged Mn_4 arrangement has been established by the EXAFS technique. Conversion of G_4–G_0 involves the adamantane to the cubane form of the Mn_xO_y type cluster [4] (Figure 3.7).

The above-mentioned cycle showed a promising pathway to couple a one-electron transfer outer sphere mechanism in stepwise oxidation by an atom transfer.

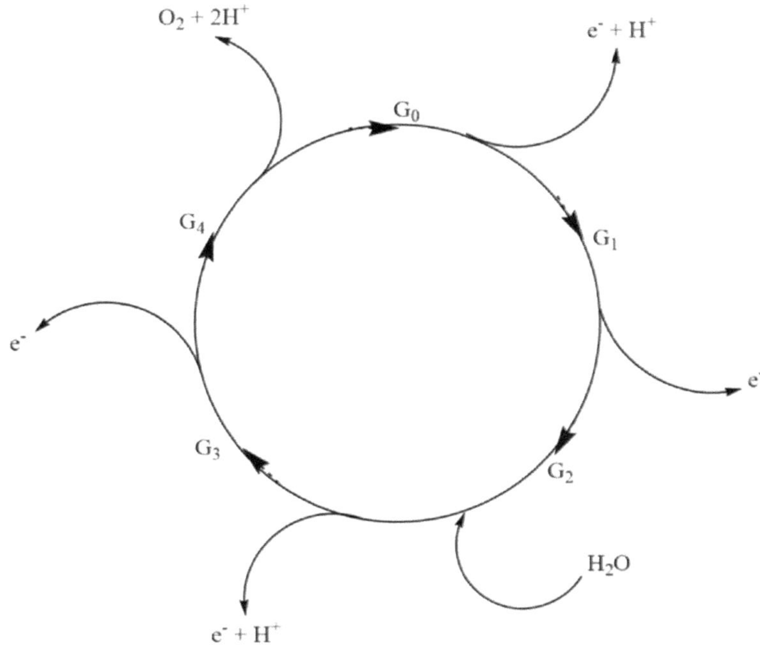

FIGURE 3.6 Consecutive steps involved in oxidation, proceeding with release of O_2 in PS II.

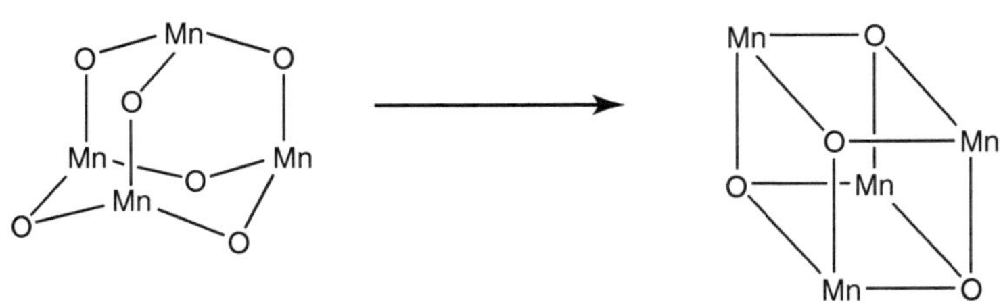

FIGURE 3.7 Conversion of adamantane to cubane form.

Basically, the reduced form of manganese cluster possesses a lower oxidation state, Mn(III) (three Mn(II) and one Mn(III), which, on oxidation, is likely to transform into three Mn(III) and one Mn(IV) (Figure 3.8).

PHOTOSYSTEM I AND II

The thylakoid membrane contains two different types of photosystems to harvest solar energy. Available chlorophyll and accessory pigments assist in capturing light and passing it to the photocenters. Photons absorbed anywhere in the harvesting zone are transferred to the P680 molecule. The cluster of pigment

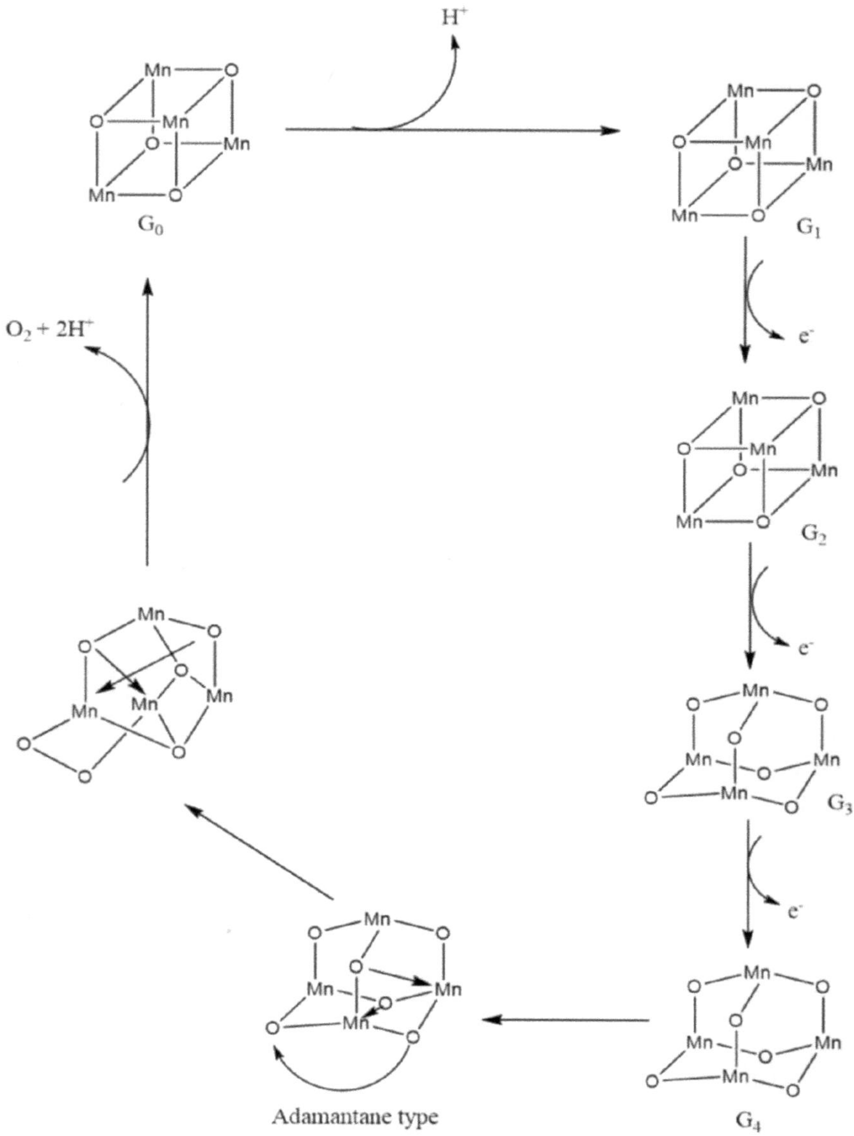

FIGURE 3.8 Mechanistic details about various manganese oxygen clusters.

molecules, along with the P680 photocenter, is collectively called photosystem II. Similarly, light absorbed at wavelengths of 700 nm or below is associated with a pigment termed photosystem II [9].

Two consecutive paths are involved in the process of photosynthesis. Photosystem I involves chlorophyll a along with other pigments for carbohydrate formation. This photosystem is also available in bacteria and was the first photosystem to evolve. Photosystem I is responsible for oxygen formation. This process requires chlorophyll a and b, which absorb at shorter wavelengths. Photosystem I contains about 200 chlorophyll a and 50 carotenoid pigments.

Eight photons are needed to complete one cycle of the reaction (two photons for each electron) from H_2O to $NADP^+$ so energy must be accumulated. In the case of PSI, energy trapping occurs. In the excited state, P_{700} loses an electron to reduce $NADP^+$. Photosystem II restores the electron to oxidized P_{700} and forms O_2 in the process. The complete reaction is represented as:

$$2H_2O + 2NADP^+ + 8\,photons \rightarrow 2NADPH + 2H^+ + O_2$$

In photosystem I, P_{700} is first photoexcited, which acts as a reductant for the Iron-Sulfur complex, and a chain process is initiated to ultimately reduce CO_2. The oxidized form of PSI (P_{700}) is not strong enough to oxidize water. The reduced form is applied for the conversion of ADP to an ATP molecule through a sequential chain reaction that includes iron-based couples and plastoquinone. The oxidized form of PSII is strong enough to oxidize water, and the reaction releases a net of four electrons.

$$2H_2O_{(l)} \rightarrow 4H^+{}_{(aq)} + O_{2(g)} + 4e^-$$

The above reaction is obtained through a redox reaction initiated by Mn complexes, which transfer the electron to P_{680}. Due to the unavailability of X-ray studies, a clear observation of the Mn cluster is not possible, but Mn cluster bonded with an oxygen atom is established. Possible changes in the Mn oxidation state are discussed in Figure 3.8.

After the absorption of a photon by photosystem I (P700), energy-rich electrons are emitted from the reaction center. Several consecutive movements involve a chain of electron carriers, destined for NADP, and protons generated by the splitting of water, which are attached to NADP to garner NADPH. The reduced moiety is used for the reduction of CO_2 in the dark reaction and is called the reducing power of the cell (Figure 3.9).

CALCIUM

Calcium is one of the major cations responsible for the structural skeleton of the body in the form of bone and shell. It exists in the insoluble form of carbonate and phosphate, such as $CaCO_3$, apatite $Ca_{10}(Po_4)_6(OH)_2$, and calcite. However, other non-structural roles are also governed by the calcium ion, e.g., messengers for hormonal action, triggering muscle reflexes, nerve signals, initiation of blood clotting, and the stabilization of protein structures. The formation constant of Ca^{2+} ion binding in muscle proteins is $5X10^4$. Other elements of the alkaline earth metal series can also bind to proteins and their order of formation is:

$$Ca^{2+} > Cd^{2+} > Sr^{2+} > Mg^{2+}$$

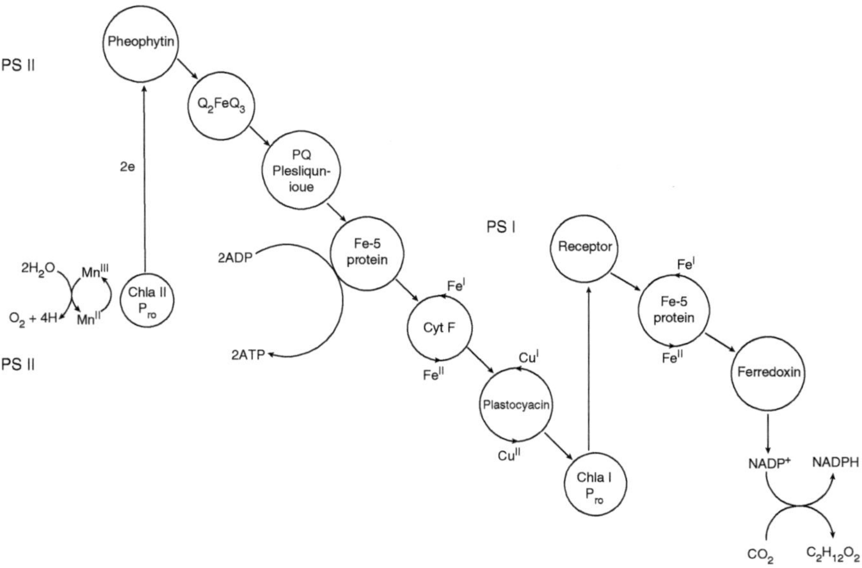

FIGURE 3.9 Depiction of photosystem I and II.

The higher formation constant of calcium can be correlated with favorable site occupancy for the hard cation to bind with a hard ligand. The hard site is tailored to chelate with Ca^{2+} ions. Most calcium-binding proteins are typically aspartate (Asp) and glutamate (Glu), both of which contain carboxylate groups in their side chains and act as hard anionic ligands [4,9]. X-ray structures reveal that Ca^{2+} binds with four carboxylate linkages, with oxygenation in a tetrahedral orientation.

Muscle contraction initiates from the nerve impulse at the neuron fiber endings. This causes Ca^{2+} to be released from the sarcoplasmic reticulum of the muscle. Consequently, the concentration of Ca^{2+} increases by 100-fold in milli-seconds due to the release from the sarcoplasmic reticulum. Upon relaxation, Ca^{2+} is again reabsorbed by the sarcoplasmic reticulum. This is a cyclic process similar to the Na^+/K^+ pump, identified as the Ca^{2+}/mg^{2+} ATPase [9].

Intake of calcium from food in the neutral phosphate form takes place in our body. The active behavior of our digestive system converts this phosphate into readily soluble acid phosphates, $CaHO_4$ and $Ca(H_2PO_4)_2$. The latter moiety is easily absorbed in the intestine and penetrates the blood plasma. The concentration of Ca^{2+} ions present in a healthy human being ranges from 0.0022 to 0.0028 moles per meter [4,9]. On average, half of the Ca^{2+} available in aqueous ions is permeable to the membrane, while the remaining amount is fixed to proteins and does not pass across the membrane.

The Ca^{2+} ions are involved not only in muscle contraction but also in the action of cardiac glycosides. An overdose of glycosides can cause cardiac arrest. Injection of K^+ and Mg^{2+} ions reduces the action of glycosides, while Ca^{2+} reverses the function. It is interesting to note that timely injection of EDTA into the arrested heart can resume the heartbeats.

Excess calcium can lead to the stone formation (lithogenesis), deposition of salts, etc. Ca^{2+} and Mg^{2+} ions are present in the cell walls of bacteria, so changes in Ca^{2+} content can assist in killing microorganisms. Introducing a strong chelating ligand, EDTA, complexed with Mg^{2+} and Ca^{2+}, can destroy the cell wall of bacteria, and kill the microorganisms.

CONCLUSIONS

An abundance of alkali and alkaline metals in the Earth's crust make these metal ions a major component of the body. These metals are critical as intracellular carriers in living organisms. They exhibit a multitude of utilities, ranging from the transportation and distribution of moieties through membranes to the storage and stabilization of proteins and various hormonal actions. Moreover, their prominence in vital biological processes, such as the maintenance of osmotic pressure, nerve signaling and photosynthesis, makes these elements essential for both plants and animals.

REFERENCES

1. D. G. Nagle, Y-D. Zhou Natural products as probes of selected targets in tumor cell biology and hypoxic signalling. In: Liu H-W, Mander L, editors. *Comprehensive Natural Products II.* Oxford: Elsevier, 2010. p. 651–683.
2. S. J. Lippard, J. M. Berg, *Principles of Bioinorganic Chemistry*, University Science Books, Mill Valley, CA, 1994, 1–585.
3. R. M. Roat-Malone, *Bioinorganic Chemistry: A Short Course*, John Wiley & Sons, Inc., Hoboken, NJ, 2002, 1–365.
4. W. Kaim, B. Schwederski, A. Klein, *Bioinorganic Chemistry: Inorganic Elements in the Chemistry of Life*, 2013, 1–426.
5. J. W. Morse, R. S. Arvidson, A. L¨uttge, Calcium carbonate formation and dissolution, *Chem. Rev.* 2007, 107, 342–381.
6. J. C. Skou, M. Esmann, The Na,K-ATPase, *J Bioenerg. Biomem.*1992, 24, 249–261.
7. Albers, R. W. Biochemical aspects of active transport, *Annu. Rev. Biochem.* 1967, 36, 727–756.
8. J. H. Kaplan, Biochemistry of Na, K-ATPase, *Annu. Rev. Biochem.* 2010, 71, 531–535.
9. D. O. Hall, K. K. Rao, *Photosynthesis* (Sixth Edition), Cambridge University Press, New York, 2001, 1–211.

4 Biological Relevance of Iron

Uptake, Electron Transfer and Transport

INTRODUCTION

Iron is the fourth most abundant element in the Earth's crust and second after aluminum as a metal. In the 3d transition metal series, iron exists in various oxidation states ranging from –II to +VI. Especially, Fe(II) and Fe(III) forms are found more often in various biological systems (Table 4.1) [1,2]. Although Fe-containing mono-oxygenase enzyme exhibit Fe(IV)/Fe(V) as intermediates in the catalytic cycle, the ubiquitous nature of iron in living systems comes from the extreme variation in the redox potential of Fe^{2+}/Fe^{3+} from −0.8 V to 0.8 V.

Both oxidation states of iron, i.e., 2^+ and 3^+, have different solubilities in water. The Fe^{3+} ion is insoluble, whereas Fe^{2+} is water-soluble [2–4]. The availability of strong chelating agents in biological systems assists solubilizing Fe^{3+} in water by forming coordination through ligands. The smaller ionic radius (0.67 Å) and higher charge categorize Fe(III) as a hard acid, which prefer, to make strong bonds with hard donor oxygen atoms like phenolate and carboxylate rather than imidazole nitrogen or thiolate sulfur atoms. Contrary to this, Fe(II) acts as an intermediate between 'hard' and 'soft' acids which enables it to coordinate with nitrogen, sulfur, and oxygen donor ligands, viz., histidine, cysteine, inorganic sulfur, and phenolate systems. Hence, the ligation of donor atoms with metal and geometry divides the functions of enzymes. For example, the aqueous group attached at the 6th position in hemoglobin is influenced by Fe^{2+} while acid-base and electron transfer reactions are dominated by Fe^{3+}. This explains the versatility of iron in biological system.

The coordination of molecular oxygen and its transport by iron-containing proteins, viz., myoglobin (Mb) and hemoglobin (Hb), is crucial to understanding the biological functions of vertebrates and many invertebrates (Table 4.1) [2]. Although the solubility of oxygen in water is low, the coordination of myoglobin and hemoglobin enhances O_2 solubility many-fold. Basically, myoglobin and hemoglobin are iron complexes coordinated through four nitrogens of a porphyrin ring. The proximal axial position is open for the coordination of oxygen/water, while the distal position is bonded with histidine nitrogen [5]. Binding of O_2 to hemoglobin is affected by the concentration of CO_2/H^+ (Bohr effect) and organic phosphates such as diphosphoglycerate (DPG) (Figure 4.1). High concentrations

DOI: 10.1201/9781003432333-4

TABLE 4.1
Distribution of Iron Containing Proteins in Adult

Protein	Mol. Wt.	Amount of Protein (g)	Amount of Fe (g)	Heme/ Non-Heme	Oxidation State	Function
Hemoglobin	64,500	750	2.60	H	Fe^{2+}	Oxygen transport
Myoglobin	17,000	40	0.13	H	Fe^{2+}	Oxygen storage
Transferrin	444,000	20	0.007	N	Fe^{3+}	Iron transport
Ferritin	–	2.4	0.52	N	Fe^{3+}	Iron storage
Hemosiderin	2,80,000	1.6	0.48	N	Fe^{3+}	Iron storage
Catalase	12,500	5.0	0.004	H	Fe^{2+}	Metabolism of H_2O_2
Cytochrome C	44,100	0.8	0.004	H	Fe^{2+}/Fe^{3+}	Terminal oxidation of H_2O_2
Peroxidase		–	–	H	Fe^{2+}/Fe^{3+}	Terminal oxidation

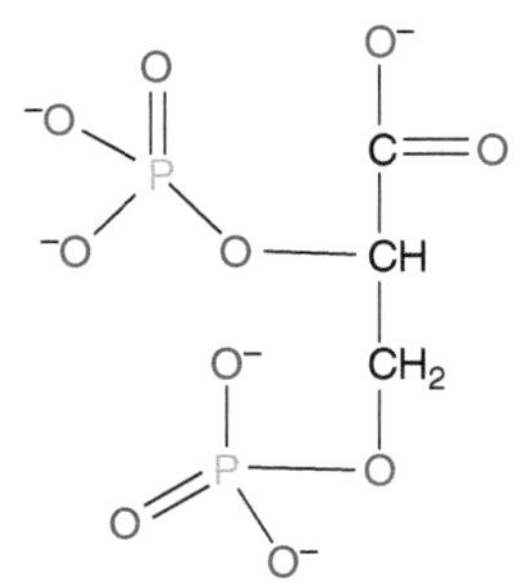

FIGURE 4.1 Diphosphoglycerate (DPG), regulator of heme-dioxygen binding.

of CO_2 and low pH decrease the O_2 affinity of Hb and consequently release the O_2 from hemoglobin [5–7]. DPG binds to hemoglobin at the distal site. Its higher concentration lowers hemoglobin's O_2 affinity and vice versa (Equation 4.1).

$$\text{Hb-DPG} + 4O_2 \rightleftharpoons \text{Hb}(O_2)_4 + \text{DPG} \tag{4.1}$$

Iron-based proteins can be classified in various ways. On the basis of functional roles, they are categorized in the following manner (Table 4.1): (i) Oxygen carriers and transport, (ii) Electron transport protein, (iii) Metal storage and transport and (iv) Catalytic (extremely large and diverse family). One can understand the diversified biochemical functions governed by iron coordinated with various ligands, supported through proteins. In this chapter, the following iron-containing proteins are covered:

1. Hemoproteins (iron-porphyrin incorporated into apoproteins involved in O_2 carriers/activators).
2. Iron-sulfur proteins (involved in electron transfer)

3. Non-heme proteins (including proteins responsible for iron storage and transport)
4. Catalytic roles of iron enzymes

BASICS OF MYOGLOBIN AND HEMOGLOBIN

Myoglobin is a monomeric globular protein present in the muscles and other tissues of vertebrates. It consists of a single polypeptide chain of 162 amino acid residues (with a molecular weight of 17.8 kDa) comprising seven α-helical (A-G) and six non-helical segments (Figure 4.2). The globin segment is connected through a chain to the imidazole ring of the histidine residue, which, as a dioxygen-binding group, is called protoporphyrin IX. The heme prosthetic center contains penta-coordinated iron with square pyramidal geometry, where the plane is defined by a pyrrole ring of the porphyrin nitrogen, along with an axial axis occupied by the imidazole ring of histidine (proximal) (Figure 4.3). In the deoxy form, myoglobin iron(II) lies 0.42 Å away from the plane of the pyrrole nitrogen, which,

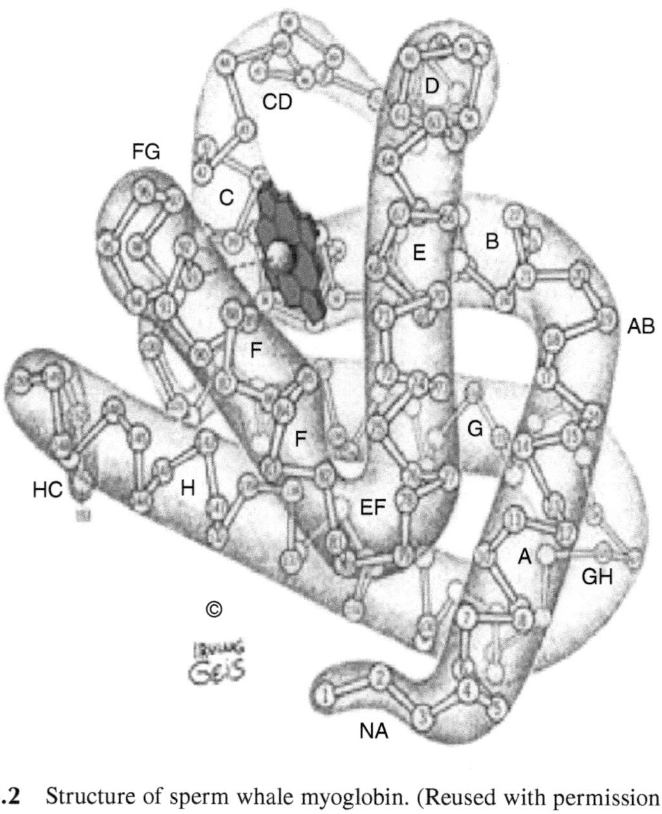

FIGURE 4.2 Structure of sperm whale myoglobin. (Reused with permission from Ref. [19], Copyright (2023) Elsevier.)

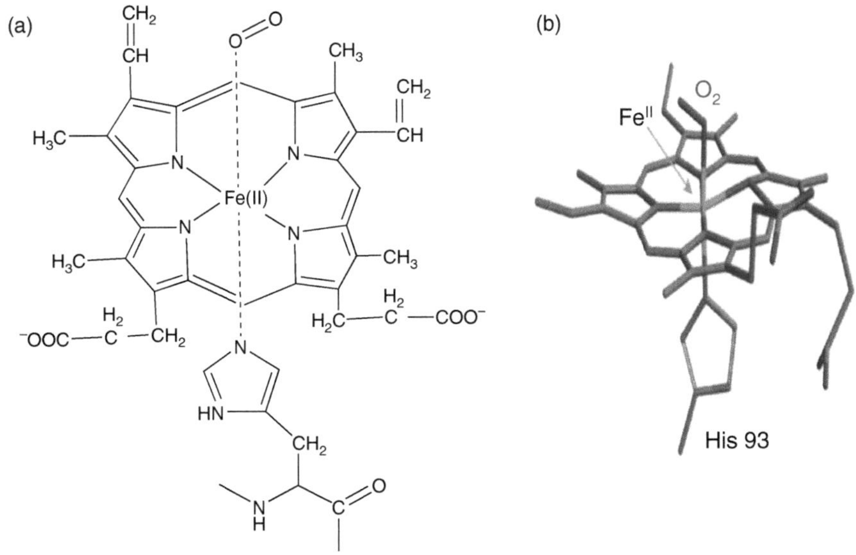

FIGURE 4.3 (a) Schematic representation of heme protein. (b) Molecular model of heme oxygen complex. (Panel (b) reused with permission from Ref. [20], Copyright (2023) Elsevier.)

upon binding with dioxygen in an end-on fashion, results in a bent configuration with an Fe-O-O bond angle of 115° [3,5].

Hemoglobin transports oxygen from the lungs to myoglobin (Mb) in tissues through blood plasma. It is a tetramer of four globular protein subunits, which are identical to Mb units. The molecular weight of hemoglobin (Hb) is approximately 64.5 kDa and consists of a tetramer of four heme groups surrounded by a polypeptide globin chain. Normally, tetrameric arrangements have two α and two β types of subunits and are represented as $\alpha_2\beta_2$. These subunits have 141 and 146 amino acid residues, respectively [2].

Both Hb and Mb bind oxygen when iron is in the 2^+ oxidation state (Table 4.2). As mentioned above, the prefixes oxy- and deoxy- refer to the oxygenated and deoxygenated forms of hemoglobin and myoglobin. Coordination of dioxygen in Hb takes place in a cooperative manner, i.e., after the attachment of one O_2 molecule, the second, third, and fourth bind more readily.

HEMOGLOBIN AND MYOGLOBIN

The stability of the Hb-O_2 complex is kinetic in nature rather than thermodynamic. To favor this stability, the following circumstances are mainly responsible:

1. Coordination of dioxygen in a bent orientation is supported by the pocket-shaped prosthetic group which prevents the formation of bridging oxo dimer.

TABLE 4.2
Comparative Properties of Various Oxygen Carriers

Characteristic	Myoglobin (Mb)	Hemoglobin (Hb)	Hemerythrin	Hemocyanin
Source	Higher animals, invertebrates	Higher animals, some invertebrates	Invertebrates	Arthropods, mollusks
Metal	Fe	Fe	Fe	Cu
O.S in deoxy form	II/d^6 (red-purple)	II/d^6 (red purple or violet)	II/d^6 (colorless)	I/d^{10} (colorless)
O.S in oxy form	II/d^6 (O_2) or III/d^5 (O_2^-) (red)	II/d^6 (O_2) or III/d^5 (O_2^-) (red)	III/d^5 (burgundy)	II/d^9 (blue)
Mol. wt (KDa)	17	65	108	400 to 2×10^4
No of subunits	1	4	8	Many
Metal: O_2	Fe:O_2	Fe:O_2	2Fe:O_2	2Cu:O_2
Coordination sphere	Porphyrin ring	Porphyrin ring	Protein side chain	Protein side chain

2. The Bent geometry of the binding mode of dioxygen is supported by σ bonding formation between the sp^2 hybridized superoxide ion and an empty d_z^2 Fe(II) orbital (Figure 4.4).
3. Back π bond formation occurs through the half-filled d_{xz} orbital of Fe(II) with the available half-filled π^* orbital of superoxide [4].
4. The heme protein lies in a hydrophobic pocket that prevents the accessibility of water molecules (Figure 4.5).

The key feature of the transformation from the T to R state is the movement of the iron atom into the porphyrin ring plane, which consequently drags the proximal histidine ligand 0.61 Å upward toward it (Figure 4.6). In the deoxygenated form of Hb, Fe(II) exists in a high-spin state (total spin=2), and its larger ionic radius cannot be accommodated in the plane of the four porphyrin rings. In contrast, the oxygenated form of Hb results in iron being in a low-spin 3^+ oxidation state ($S=1/2$). During the switch from Fe(II) to Fe(III), the ionic radius of iron decreases, allowing it to fit into the cavity of the porphyrin ring [2,4].

The protein pocket favors the orientation of the oxygen molecule to coordinate with the Fe(II) center in a bent fashion, having an Fe-O-O bond angle of 115°. Binding of dioxygen oxidizes Fe(II) to the Fe(III) state, and dioxygen becomes a superoxide ion (O^{2-}). The existence of the Fe(III)-O^{2-} species is well supported by Fe-O, Fe-O-O, and O-O bond distances, bond angles and their bond order as determined through X-ray crystallography, infrared, and resonance Raman spectroscopy, respectively (Tables 4.3–4.5). The singlet spin state ($S=0$) of the Fe(III)-O_2^- species indicates magnetic coupling of the Fe(III) ions unpaired electrons with those of the superoxide ion [4,6–9].

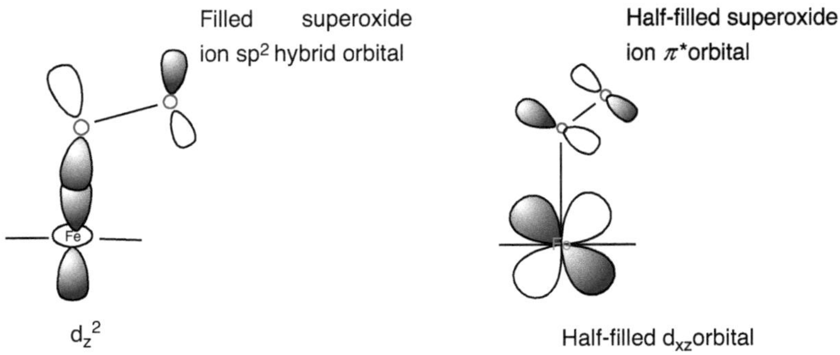

FIGURE 4.4 Bonding mode of O_2^- ion to Fe(III) in hemoglobin and myoglobin. (Reused with permission from Ref. [21], Copyright (2023) American Chemical Society.)

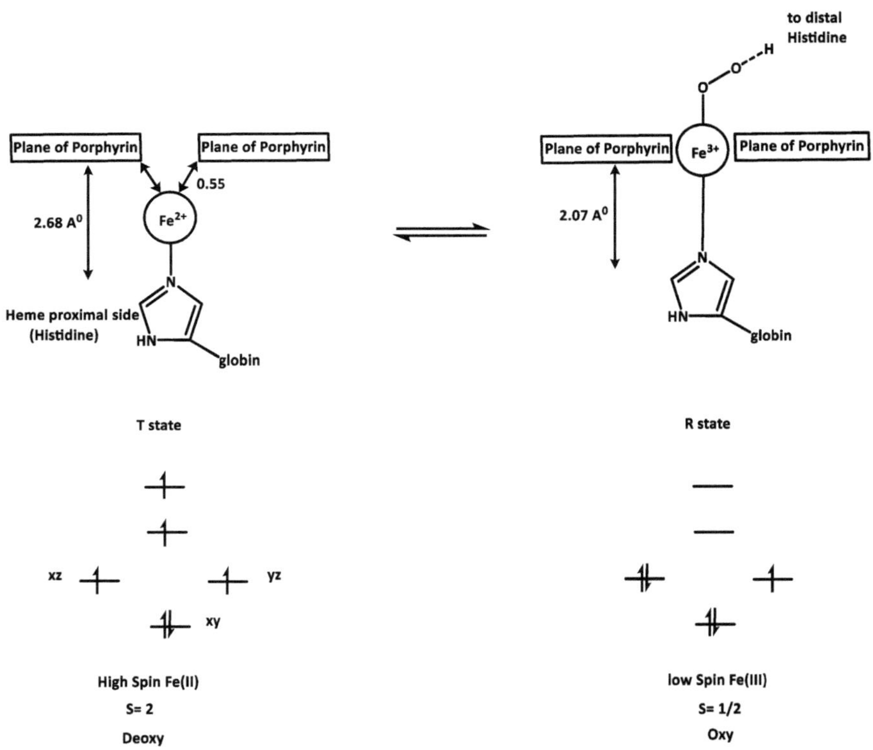

FIGURE 4.5 Illustration of confirmation and change in spin state during transformation of T and R states.

α and β refer to the protein chains in hemoglobin.

Φ is defined as the angle between the plane of the axial base (histidine) and the plane of the defined metal (M ligands of the porphyrin ring).

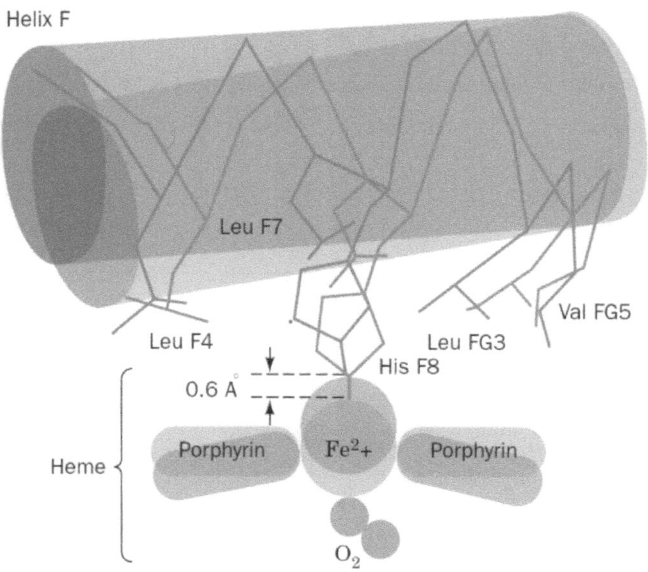

FIGURE 4.6 Sketch representing transformation of T and R states. (Reused with permission from Ref. [19], Copyright (2023) Elsevier.)

TABLE 4.3
Structural Parameter of Oxyhemoglobin

Parameter	MbO_2	HbO_2 (α)	HbO_2 (β)
Fe-N (porphyrin) (Å)	1.95 (6)	1.99 (5)	1.96 (6)
Fe-N (proximal histidine) (Å)	2.07 (6)	1.94 (9)	2.07 (9)
Fe-porphyrin (Å)	0.18 (6)	0.121 (8)	–0.11(8)
Fe-O (Å)	1.83 (6)	1.66 (8)	1.87 (3)
O-O (Å)	1.22 (6)	–	–
Fe-O-O	115(5)	153 (7)	159 (12)
Φ°	1	11	27
Electronic spectra			
π ……>π* (α and β bands)	–	400–600 nm	–
Vibrational spectra	–	1,105 cm^{-1}	–
O-O (stretching)			

HEMERYTHRIN

Hemerythrin is a non-heme protein found in certain marine invertebrates. It occurs intracellularly in sipunculids, brachiopods and polychaetes. Unlike two other respiratory proteins, i.e., hemoglobin and hemocyanin, hemerythrin does not contain a heme group [2,4]. Despite the functional similarities among these

TABLE 4.4
Structural Parameter of O_2 Molecule and Ions

Species	Bond Order	Bond Length (Å)	$\upsilon_{(o\text{-}o)}$ cm^{-1}
O_2	2.0	1.21	1,560
O_2^- (superoxide)	1.5	1.28	1,150–1,100
O_2^{2-} (peroxide)	1.0	1.49	850–740

TABLE 4.5
Structural Parameter of Deoxy Heme Protein (Myoglobin and Hemoglobin)

Bond Length Å/Bond Angle (°)	Mb	Hb_α	Hb_β
Fe-N (porphyrin)	2.03 (10)	2.08(3)	2.05(3)
Fe-N (proximal histidine)	2.22	2.16(6)	2.09(6)
Fe-porphyrin	0.42	0.40(5)	
Φ	19	18 (1)	

three oxygen carriers—hemoglobin, hemocyanin, and hemerythrin—they show dissimilarities with respect to their molecular structure, the molecular nature of the oxygen-binding site, etc. A comparative perspective on these proteins is mentioned in Table 4.2.

A general consideration related to the 'heme' prefix indicates the presence of an iron-porphyrin ring, whereas the word 'heme' was originally adopted from the Greek word for blood. Hemocyanin and hemerythrin do not contain a porphyrin ring: however, their oxygenated states show blue and violet appearances, respectively.

Hemerythrin exists in an octameric form with a molecular weight in the range of 100–110kDa. In a few species, dimeric, trimeric, and tetrameric forms of the pigment are also present. The quaternary structure of octameric hemerythrin from *P. gondii* is composed of two layers of identical or quasi-identical four subunits arranged in an end-to-side fashion [2–4]. The shape of the molecule appears as a square doughnut. Hydrogen bonding and electrostatic interactions between both layers hold the structure together to constrain the octamer. The monomeric form of hemerythrin is called myohemerythrin, which is responsible for oxygen storage.

Active Site

The stoichiometric ratio of 2:1 for Fe:O_2 in hemerythrin provides an indication of the bridging oxygen flanked between two iron atoms. High-resolution crystallographic analysis supports the presence of the imidazole ring of

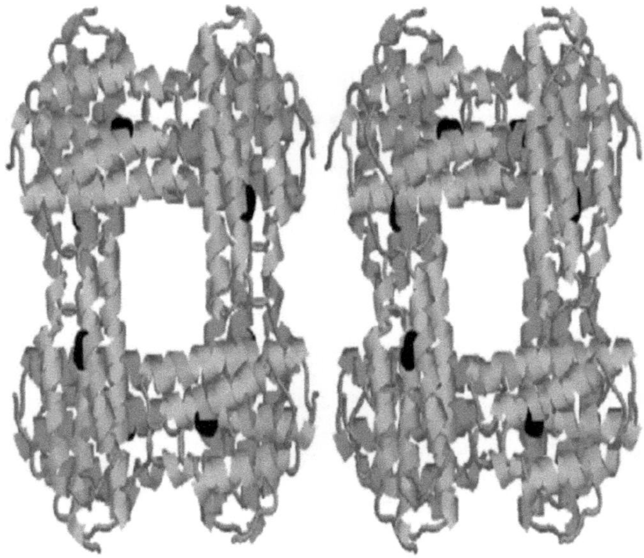

FIGURE 4.7a Quaternary structure of an octameric hemerythrin. (Reused with per-mission from Ref. [22], Copyright (2023) Elsevier.)

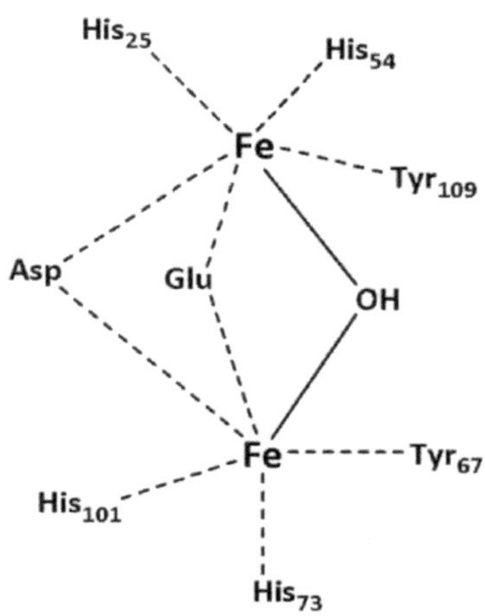

FIGURE 4.7b Representation of hemerythrin.

Histidines 25 and 54, and Tyrosine 109, coordinated to one iron, whereas Histidines 72,101 and either Histidine 77 or Tyrosine 67, are attached to the second one. Glutamic acid-58 and Aspartic acid-106 act as bridging ligands attached to both iron atoms [10].

The Fe-Fe distance at the active site of hemerythrin's probable structure is 3.44 ± 0.05 Å, which is compatible with an oxo-bridged structure [7,10].

TABLE 4.6
Structural and Spectroscopic Properties of Hemerythrin

Physical Parameters	Oxyhemerythrin	Deoxyhemerythrin
Fe-O-Fe	128	128
Fe-μO	1.82	1.98
Fe-----Fe distance (Å)	3.24	3.57
Resonance Raman spectra, vibration band	$848\,cm^{-1}$ (peroxide presence)	–
Electronic spectra	330, 360°	
LMCT (peroxide to Fe(III))		
Vibration spectra		
$Fe\text{-}O\text{-}Fe_{sym}$	486	–
$Fe\text{-}O\text{-}Fe_{asym}$	757	–
O.S.	III	II

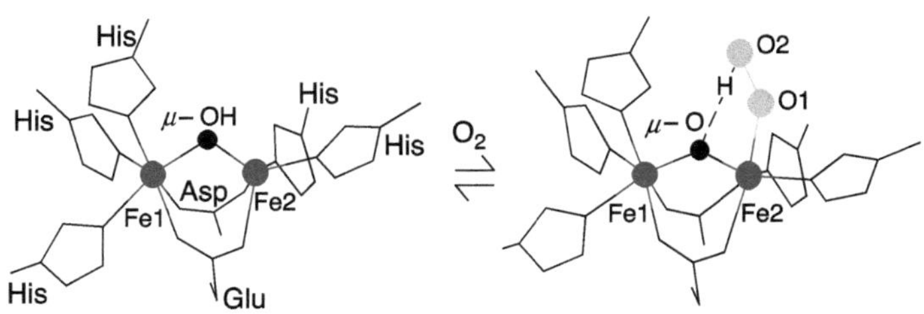

FIGURE 4.8 Structural changes in oxy and deoxy hemerythrin. (Reused with permission from Ref. [22], Copyright (2023) Elsevier.)

The shorter bond distance $Fe\text{-}O_{oxo\,(bridge)}$ is 1.80 Å, with a bridging Fe-O-Fe angle of 145° and elongated bond lengths of 2.05–2.25 Å for Fe-O and Fe-N terminal bonds, which support the presence of a high-spin d^5 core center in the oxidized form of the pigment [10] (Table 4.6). Essentially, each subunit of hemerythrin consists of two Fe(II) centers, in which one iron atom is penta-coordinated, and the second iron has six coordination sites to attain octahedral geometry (Figure 4.8).

Mechanism

Although hemerythrin is analogous to hemoglobin, it is not homologous. In the deoxygenated form of hemerythrin, the iron pair exists in a high-spin Fe(II) state, with similar environments. The addition of oxygen results in oxyhemerythrin, in which both irons are oxidized to high-spin Fe(III) ions, bridged by an oxygen molecule. The bound dioxygen is initially reduced to peroxide (O_2^{2-})

FIGURE 4.9 Change in the oxidation state of iron during oxygenation of hemerythrin.

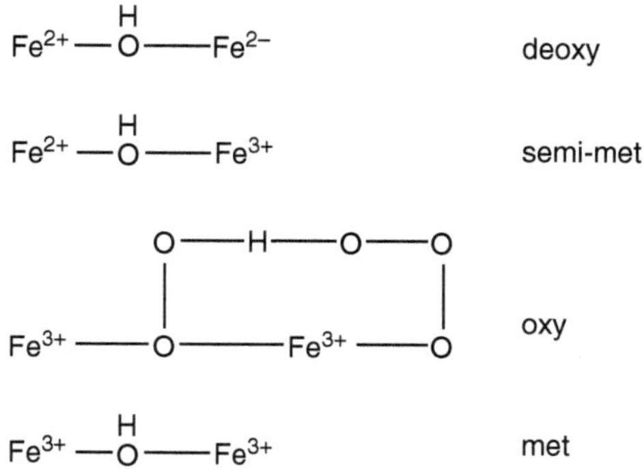

FIGURE 4.10 Various classes of hemerythrin depending upon oxygenation.

(Figures 4.9 and 4.10). Spectroscopic and magnetic techniques have shown that in oxy-hemerythrin, both Fe(III) atoms are coupled antiferromagnetically [5].

The oxidation state assigned to oxygen, O_2^{2-}, emerges by implication, with two electrons being given up by 2Fe(II) of deoxy-hemerythrin, resulting in 2Fe(III) in oxy-hemerythrin. The attachment of oxygen to hemerythrin does not follow a cooperative effect.

Structural Changes

Deoxy-hemerythrin contains two ferrous ions bridged with hydroxyl groups. Among the two irons, one is hexacoordinated, whereas the second has a penta-coordination nature. The role of the hydroxyl group is not quite clear; however, it is believed that the hydroxyl group acts as a source to provide protons to the incoming dioxygen molecule (Figure 4.11). The oxygen molecule attaches to the penta-coordinated Fe(II) center at the available vacant position. Then, electrons flow from Fe(II) to generate Fe(III) with the bound peroxide.

RELATIVE AFFINITY AND EFFICIENCY OF HEMOGLOBIN AND MYOGLOBIN FOR OXYGEN

The affinity of oxygen toward hemoglobin is less compared to its structural analogue myoglobin. This interesting feature makes hemoglobin a more efficient carrier for oxygen in vertebrates. Another aspect to observe in this property is that

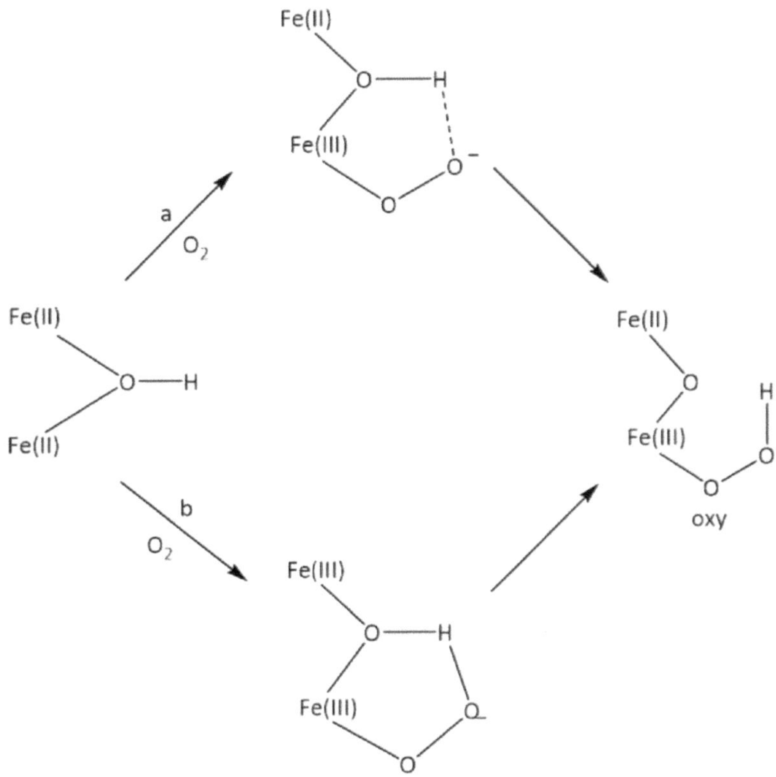

FIGURE 4.11 Mechanistic pathway for O_2 transfer in hemerythrin.

the lesser inclination of any pigment to bind oxygen, in turn, assists in the facile release of the binding molecule whenever it is needed. However, a higher affinity of a protein for oxygen makes the release of oxygen harder, whereas the lower affinity of hemoglobin for oxygen assists in the easier release of oxygen to various organs. On the contrary, the higher extent of binding of oxygen to myoglobin makes it less inclined to release oxygen. We cannot overlook a more suitable reason for hemoglobin to be chosen by the body to distribute oxygen; however, myoglobin also serves this purpose, specifically for muscle cells.

Affinity of O_2 to Hb and Mb is a function of oxygen partial pressure (Figure 4.12). The hyperbolic curve belongs to myoglobin, where as a sigmoidal nature is observed in the case of hemoglobin. The difference in the oxygen dissociation curves among both pigments can be very well understood through the cooperative effect. The details of this effect are discussed later. At the lungs, the saturation of oxygen with hemoglobin is 100% at a partial pressure of 100 torr (13 kpa). Moving from the lungs to the muscles, the saturation fraction decreases with the decreasing partial pressure. At 50% saturation, the calculated partial pressure is 26 torr (calculated). At the lower end of the curve, i.e., 15 torr, one can easily notice the higher affinity of myoglobin with respect to hemoglobin [8].

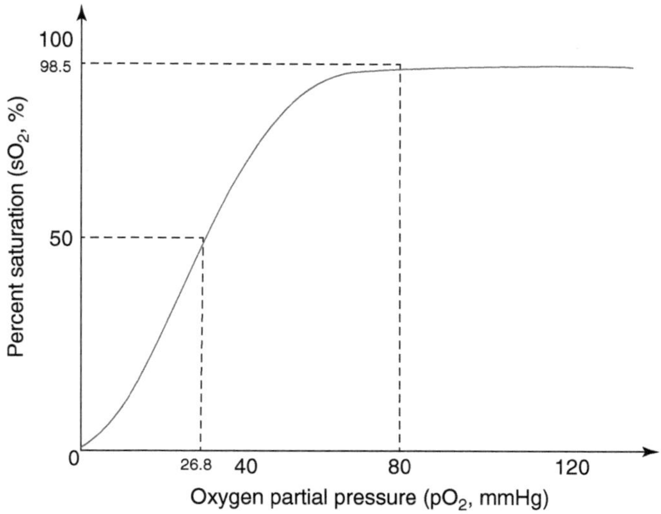

FIGURE 4.12 Oxygen binding curve of hemoglobin.

COOPERATIVE EFFECT

In many proteins, the binding of the first ligand to the protein changes the binding affinity of the second ligand. This behavior can be explained by the cooperative mechanism, which is an important regulatory phenomenon in biochemistry. To understand the nature of cooperative binding, let us take a general reaction where two ligands bind with the immunoglobulin (protein) in the following manner:

$$P + L \rightarrow PL \rightarrow PL2$$

K1 = Affinity constant for step 1; K2 = Affinity constant for step 2

If the value of K_2 is greater than K_1, positive cooperativity will be present, whereas the reverse scenario falls under negative cooperativity. The nature of cooperative binding can also be determined by the binding curve plotted between saturation fraction (Ⱦ) and ligand concentration [10]. The hyperbolic origin of the binding curve corresponds to non-cooperativity, while sigmoidal behavior belongs to positive cooperative binding.

Quantitative prediction of the above-mentioned graph shows that, in the case of positive cooperativity, a small change in the ligand concentration results in drastic changes in the ligand-loaded protein (Figure 4.13). However, negative cooperativity can be noticed even if a large change in the concentration of the ligand makes an equivalent change in the ligand-bound protein [10].

COOPERATIVITY EFFECT IN HEMOGLOBIN

In hemoglobin, the sigmoidal shape of the oxygen dissociation curve reflects the positive cooperativity of oxygen binding. This means that hemoglobin transmits the intracellular signal to its own binding site to attain maximum affinity for

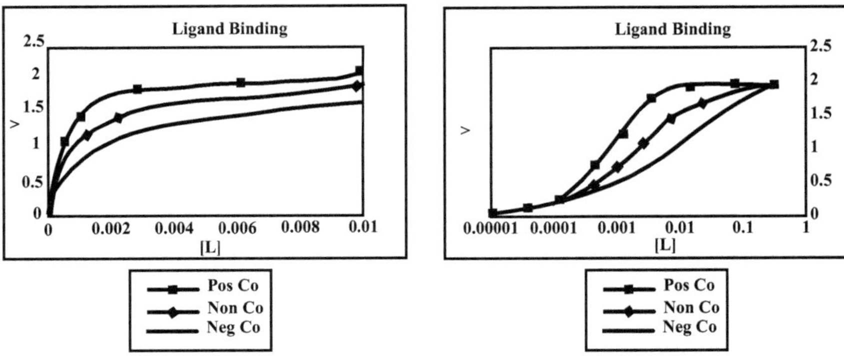

FIGURE 4.13 Effect of ligand concentration on positive and negative cooperativity.

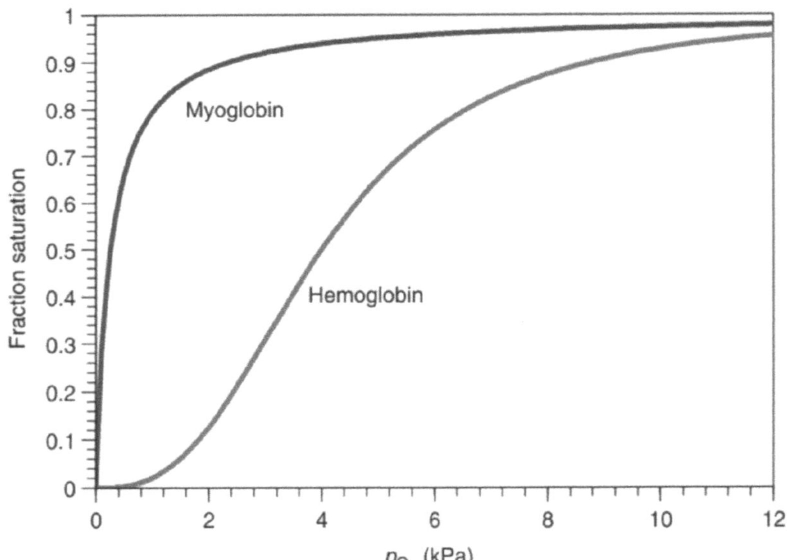

FIGURE 4.14 Comparative oxygen binding curve of hemoglobin and myoglobin. (Reused from Ref. [23].)

oxygen molecules. Hemoglobin exists in tetrameric form, and once the monomer of hemoglobin binds with oxygen, it initiates an alarm for the other subunits of Hb to start the binding process (Figure 4.14). This means that the more oxygen is coordinated with hemoglobin, the more the affinity of hemoglobin will increase. Since the binding of one molecule (O_2) affects the binding of the successive molecule, such an interaction is called homotropic allosteric interaction.

To understand the cooperative effect in the case of hemoglobin, let us take a modified example of Voet [10], comparing the cooperativity of Hb and oxygen to a Mumbai local train compartment and passengers, respectively.

Suppose only four seats are vacant in the compartment. The passengers are compelled to occupy a partially filled compartment seat. At the final destination, a few passengers depart, and the remaining fellow passengers are more likely to vacate if the compartment mates have already left [11–14].

This analogy will help to explain why the upper and lower ends of the oxygen dissociation curve are important. At the upper end of the curve (peak hours), passengers find it difficult to board since the compartment is full, whereas at the lower end (off-peak hours), passengers dislike boarding since the compartment is empty.

The hyperbolic nature of the dissociation curve in a protein like myoglobin exhibits the non-cooperative phenomenon, which can be compared with passengers in the first- or second-class compartment of an express train, where no one bothers about the presence or absence of fellow passengers. Hope this approach will be helpful in perceiving the cooperative interaction between ligand and protein.

HILL EQUATION

The very first mathematical equation to explain the binding of ligands to macromolecules, as a function of ligand concentration, was given by Archibald Vivian Hill in 1910. He received the Nobel Prize in Physiology for this work. The Hill equation was primarily formulated to describe the oxygen dissociation curve of hemoglobin [11–13]. It is also helpful in quantifying the degree of cooperativity with the help of the Hill coefficient (n), which is the slope of the Hill plot ($\log \theta$ vs $\log [L]$). The Hill equation is expressed by:

$$\theta = \frac{[\mathrm{L}]^n}{K_d + [\mathrm{L}]^n}$$

$$= \frac{[\mathrm{L}]^n}{(K_a)^n + [\mathrm{L}]^n}$$

$$\theta = \frac{1}{1 + \left(\frac{K_A}{[\mathrm{L}]}\right)^n}$$

$$\theta = \frac{1}{1 + \left(\frac{K_A}{[\mathrm{L}]}\right)^n}$$

θ=Fraction of receptor protein concentration bound by the ligand
[L]=Total ligand concentration
K_d=Equilibrium constant for dissociation
K_A=Ligand concentration producing half occupation
n=Hill coefficient

$$1 + \left(\frac{K_A}{\mathrm{L}}\right)^n = \frac{1}{\theta}$$

$$\left(\frac{K_A}{\mathrm{L}}\right)^n = \frac{1}{\theta} - 1$$

$$\left(\frac{K_A}{\mathrm{L}}\right)^n = \left(\frac{1-\theta}{\theta}\right)$$

Reciprocate the equation on both sides,

$$\left(\frac{\mathrm{L}}{K_A}\right)^n = \left(\frac{\theta}{1-\theta}\right)$$

$$\frac{[\mathrm{L}]^n}{(K_A)^n} = \left(\frac{\theta}{1-\theta}\right)$$

$$\boxed{\log\left(\frac{\theta}{1-\theta}\right) = n\log[\mathrm{L}] - \log K_A}$$

Taking the logarithm of both sides of equation:-

$$(K_A)^n = K_d$$

$$\boxed{\log\left(\frac{\theta}{1-\theta}\right) = n\log[\mathrm{L}] - \log K_d}$$

A plot of $\log\left(\frac{\theta}{1-\theta}\right)$ vs log[L] results in a linear plot, which is called the Hill plot (Figure 4.15). the slope of the Hill plot corresponds to the Hill coefficient, n. The value of n explains the nature of cooperativity of ligand binding in following way:

If slope

$n > 1$, binding shows positive cooperativity

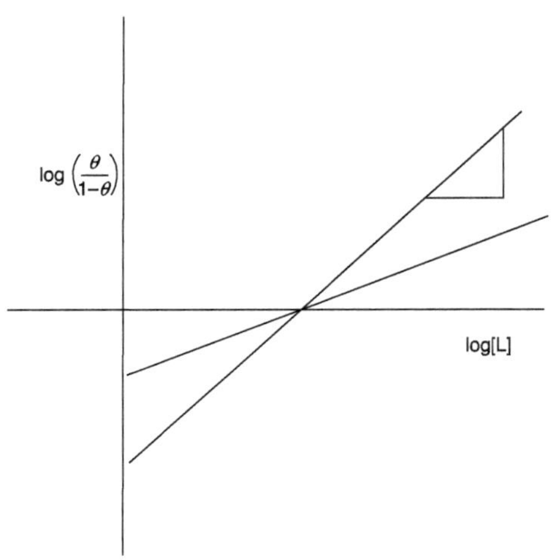

FIGURE 4.15 A plot between $\log\left(\frac{\theta}{1-\theta}\right)$ and $\log[L]$.

$n<1$, binding shows negative cooperativity
$n=1$, Non cooperative binding
At $n=1$, $K_D=K_A=K_M$, the Michaelis Menten constants

Derivation of Hill Equation

The binding of ligands to the protein can be represented by the following reaction:

$$\mathrm{P}+n\mathrm{L} \underset{K_d}{\overset{k_a}{\rightleftharpoons}} \mathrm{PL}n$$

As per collision theory, the equilibrium constant $K_{\mathrm{eq}}=$

As per the collision theory, equilibrium constant $K_{\mathrm{eq}} = \dfrac{K_d}{K_a}$

$$\mathrm{P}+n\mathrm{L} \overset{K_a}{\rightarrow} \mathrm{PL}_n \;\&\; \mathrm{PL}n \overset{K_d}{\rightarrow} \mathrm{P} + n\mathrm{L}$$

At equilibrium

$$\frac{d[\mathrm{L}]}{dt} = -K_a\,[\mathrm{P}][\mathrm{L}]^n + K_d[\mathrm{PL}_n]=0$$

$$+K_a[\mathrm{P}][\mathrm{L}]^n = +K_d[\mathrm{PL}_n]$$

$$[\mathrm{PL}_n]=\left(\frac{K_a}{K_d}\right)[\mathrm{P}]\,[\mathrm{L}]^n \tag{4.2}$$

As we know that:

$$\theta = \frac{\text{Occupied receptor}}{\text{Total receptor}} = \frac{\text{Binding sites occupied}}{\text{Total binding sites}}$$

θ is the value of the concentration of occupied receptor to total receptor concentration

$$= \frac{[\mathrm{PL}_n]}{[\mathrm{P}]+[\mathrm{PL}_n]} \tag{4.3}$$

Substitute the value of the Equations (4.2)–(4.3)

$$= \frac{\left(\frac{K_a}{K_d}\right)[\mathrm{P}]\,[\mathrm{L}]^n}{[\mathrm{P}]+\left(\frac{K_a}{K_d}\right)[\mathrm{P}][\mathrm{L}]^n} \tag{4.4}$$

Divide the equation through 'P',

$$= \frac{\left(K_a/K_d\right)[\mathrm{L}]^n}{1+\left(K_a/K_d\right)[\mathrm{L}]^n}$$

Binding to Site

$$P + L \rightleftharpoons PL$$

K_{eq}=Equilibrium constant, also known as the association constant or affinity constant

$$K_{eq} = \frac{[PL]}{[P][L]}$$

[PL]=Concentration of the protein-ligand complex
[P]=Concentration of the protein
[L]=Concentration of the free ligand
Dissociation constant

$$K_D = \frac{[P][L]}{[PL]}$$

$$\boxed{K_D = \frac{1}{K_{eq}}}$$

$$K_D = \frac{K_d}{K_a} = \text{Microscopic dissociation constant}$$

$$\theta = \frac{K_d[L]^n}{1 + K_d[L]^n}$$

$$\boxed{\theta = \frac{[L]^n}{K_D + [L]^n}}$$

The above equation is known as the Hill equation.
Implementation of Hill Equation to Myoglobin:

$$Mb + O_2 \leftrightarrow MbO_2 \qquad K_{eq} = \frac{[MbO_2]}{[Mb][O_2]}$$

Θ is defined as fractional saturation, i.e., the fraction of protein (Mb) molecules that are saturated with oxygen.

$$\theta = \frac{[MbO_2]}{[Mb] + [MbO_2]}$$

$$K_{eq} = \frac{[MbO_2]K_{eq}}{[Mb] + [Mb][O_2]K_{eq}} \tag{4.5}$$

$$K_{eq} = \frac{K_{eq}O_2}{1 + K_{eq}O_2}$$

$$\boxed{\theta = \frac{[O_2\ ]}{1 + K_{eq}O_2}}$$

When myoglobin is half-saturated with oxygen, $[MbO_2] = [Mb]$.

We can use the term P_{50} or $P_{1/2}$, and the corresponding constant will be K_p. Equation (4.5) can be written as:

$$\frac{1}{2} = \frac{P[O_2]½}{K_P + P[O_2]½}$$

$$K_P + P[O_2]½ = 2P[O_2]½$$

$$\boxed{K_P = P[O_2]½}$$

Similarly, we can implement the above equation for hemoglobin (Figure 4.16):

$$Hb + O_2 \rightleftharpoons Hb(O_2)_4$$

$$\theta = \frac{[P(O_2)]^4}{[P(O_2)]^4 + [P_{1/2}(O_2)]^4}$$

$$\boxed{\log\left(\frac{\theta}{1-\theta}\right) = n \log PO_2 - n \log P_{1/2}O_2}$$

Θ = Oxygen saturation
$P(O_2)$ = Partial pressure of oxygen
$P_{1/2}(O_2)$ = Partial pressure at which half the iron will be oxygenated.

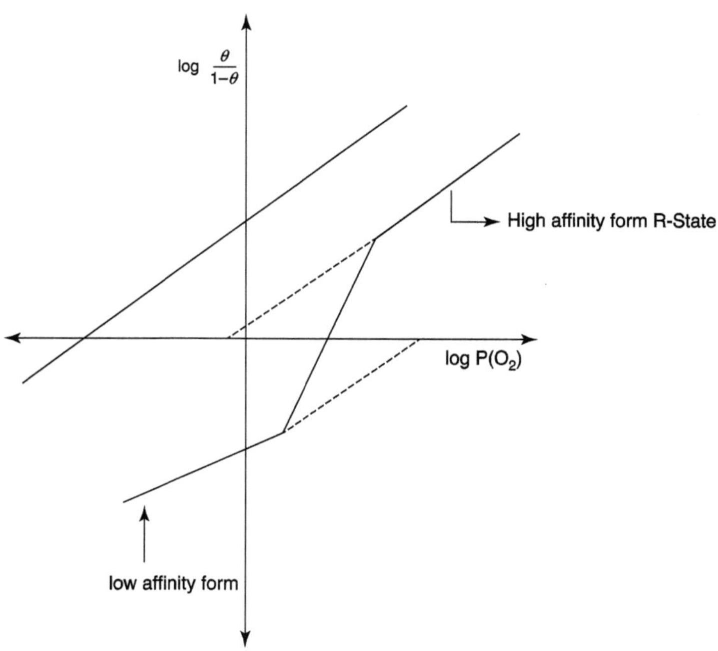

FIGURE 4.16 Illustration of T and R states with the plot between $\log\left(\frac{\theta}{1-\theta}\right)$ and $\log[L]$.

BOHR EFFECT

Oxygen cooperatively and reversibly binds with hemoglobin. Certain changes in physiological conditions such as pH, H^+ concentration, CO_2 levels, temperature, and the concentration of 2, 3- diphosphoglycerate (2, 3-DPG) affect the affinity of oxygen toward hemoglobin. In this context, the correlation of O_2 affinity with respect to pH was explained by Danish physiologist Christian Bohr in 1904 [12,13]. The Bohr effect describes that hemoglobin binds to oxygen with less affinity at low pH.

During cell respiration, an increase in metabolic activity results in the production of CO_2 as a byproduct. The increase in P_{CO_2} in tissues leads to an increased H^+ concentration, represented as a decrease in pH value due to acidolysis. This, in turn, weakens the affinity of hemoglobin for oxygen. The hemoglobin dissociation curve shifts rightward, as hemoglobin unloads O_2 from its binding site at higher PO_2. At lower pH, the availability of abundant H^+ concentration results in their association with amino acids in hemoglobin, causing conformational changes in protein folding. This ultimately reduces the affinity for O_2 binding. This configurational shift in hemoglobin is called the T (taut) form.

MECHANISM

Hemoglobin displays two different roles, the taut (T) form and the relaxed (R) form. As mentioned earlier, the T state leads to low affinity for oxygen molecules, whereas the R form is responsible for oxygen binding (higher affinity). Normally, the T form is favored by two salt bridges formed by three amino acids. One salt bridge is formed by the interaction between the β-His-146 (carboxylate group) and the terminal α-Lysine-40. The second salt bridge originates due to the connection between an additional available hydrogen on histidine and the negatively charged aspartate-94 (Figure 4.17). Thus, at lower pH (higher proton concentration), this salt bridge formation is relatively easier. The better salt bridge formation supports the formation of a strong T state, which results in a low affinity for oxygen binding to hemoglobin. The reverse nature has been observed under reverse conditions i.e., at higher pH. At lower hydrogen ion concentrations, the formation of salt bridges is more difficult, resulting in a less stable T state (Figure 4.18). This latter state is more inclined toward oxygen binding, leading to a higher affinity for oxygen binding to hemoglobin. The dual tendency of hemoglobin toward oxygen binding, depending on pH conditions [10–14], indicates the allosteric effect of hydrogen ion concentration.

Cellular respiration releases CO_2 and water as byproducts. The presence of the enzyme carbonic anhydrase catalyzes the reaction of CO_2 and H_2O to form carbonic acid. To maintain biochemical equilibrium, carbonic acid dissociates into H^+ and HCO_3^- ions (Figure 4.19). This release of hydrogen ions decreases the pH in the blood, which, in turn, decreases oxygen affinity toward hemoglobin. These reactions occur in peripheral tissues, promoting the dissociation of oxygen from hemoglobin to the tissues and facilitating oxygen loading in the lungs.

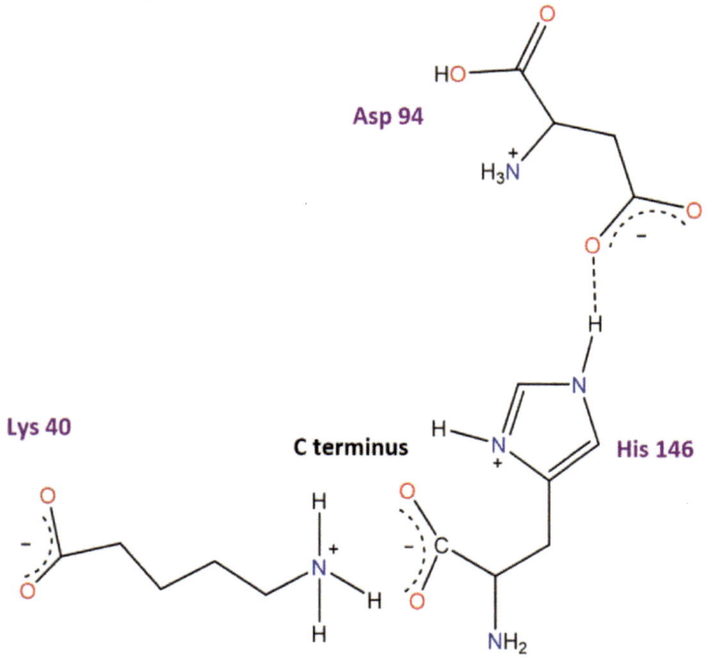

FIGURE 4.17 Diagrammatic representation of salt bridge formation in hemoglobin.

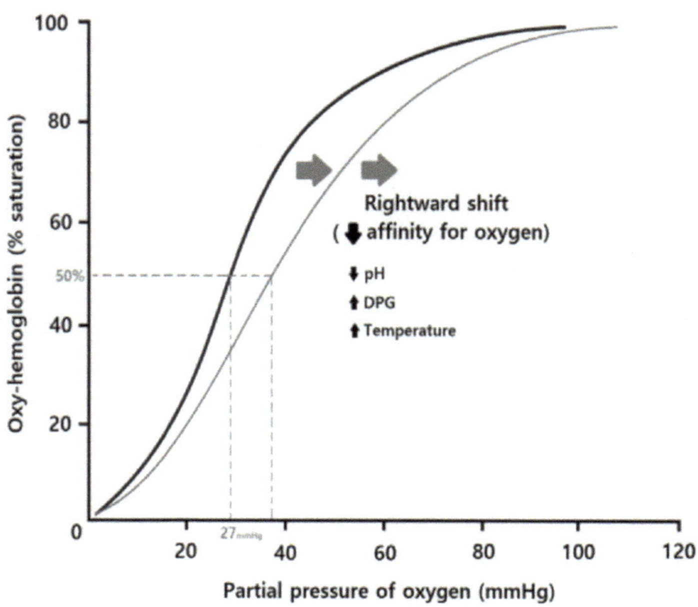

FIGURE 4.18 Dissociation curve of oxygen from hemoglobin depending on pH value. (Reused from Ref. [24].)

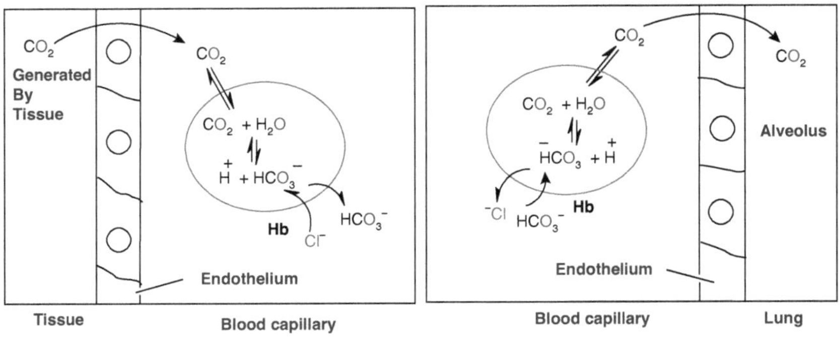

FIGURE 4.19 Allosteric effect of carbon dioxide on hemoglobin.

In this oxygen exchange process, hemoglobin acts as a buffering agent to maintain the binding of H^+ ions. The increased HCO_3^- species move down their concentration gradient by diffusing out from red blood cells [15,17]. To maintain charge neutrality across the RBC membrane, chloride ions enter the cell. This buffering process is known as the Haldane effect. Overall, carbon dioxide acts as another allosteric inhibitor for hemoglobin.

EFFECT OF CARBON MONOXIDE

The presence of carbon dioxide leads to the more prominent unloading of oxygen, whereas carbon monoxide has the opposite effect. Carbon monoxide binds to hemoglobin and myoglobin with 200 and 25 times greater affinity than oxygen, respectively. Even though the selective binding of dioxygen over carbon monoxide occurs due to steric repulsion caused by amino acid residues present on the distal side of the porphyrin ring, preferential hydrogen bonding favors dioxygen rather than carbon monoxide. The dioxygen molecule prefers a bent geometry to coordinate with the iron metal center of the heme group, which is favored through the hydrogen bonding to the distal histidine [14,16]. In contrast, carbon monoxide prefers a linear binding perpendicular to the metal plane. This not only prevents the formation of hydrogen bonds but also experiences steric repulsion caused by neighboring amino acids.

IRON SULPHUR PROTEIN

Billions of years ago, during the primitive era, the environment was reductive in nature with an abundant bioavailability of iron and sulfur elements. Consequently, iron-sulfur proteins were among the first catalysts to be profusely present in nature. The International Union of Biochemistry classified iron-sulfur proteins into two categories: simple and complex iron-sulfur proteins. Simple Fe-S proteins are those in which Fe-S is the only prosthetic group. In contrast, complex

Fe-S proteins have additional prosthetic groups such as flavins, Mo-W cofactors, flavin-heme, and others [17]. Essentially, complex proteins contain iron bound to sulfur atoms linked to the polypeptide chain, connected to thiol of cysteine amino acids or with inorganic sulfides and cysteine thiol ligand systems. The utility of Fe-S proteins is not limited to electron transfer; they also function to bind oxygen and nitrogen atoms of organic substrates in enzymes with redox and non-redox roles (e.g., nitrogenase and aconitase). Unsurprisingly, the broader utility of Fe-S proteins has led to discovery of 160 Fe-S enzymes to date. Overall, four basic core structures have been well characterized and are outlined as follows.

RUBREDOXINS

Rubredoxins are the simplest redox iron-sulfur proteins, found in bacteria. They contain one iron center coordinated to four cysteine amino acid residues in a tetrahedral fashion (Figure 4.20). During electron transfer, the colorless Fe(II) state ($S=2$) is oxidized to the red Fe(III) form without significant changes in Fe-S distances, which range between 224–233 pm. The appearance of the red color results from ligand-to-metal charge transfer from the thiolate ligand system to the iron(III) center. The biological role of rubredoxin is not precisely certain, through some rubredoxins participate in fatty acid hydroxylation. Rubredoxin acts as a single-electron donor-acceptor protein, often abbreviated as Fe_1S_0.

$$Fe^{II} \rightarrow Fe^{III} + e^-$$

$$S = 2 \quad S = 5/2$$

$$E_0 = 0 - 100 \text{ mv}$$

[2Fe-2S] proteins are also known as plant ferredoxins, which are obtained from spinach leaves. Unlike rubredoxins, the redox center consists of an equal number of iron and sulfur atoms represented as (2Fe-2S) [15–17].

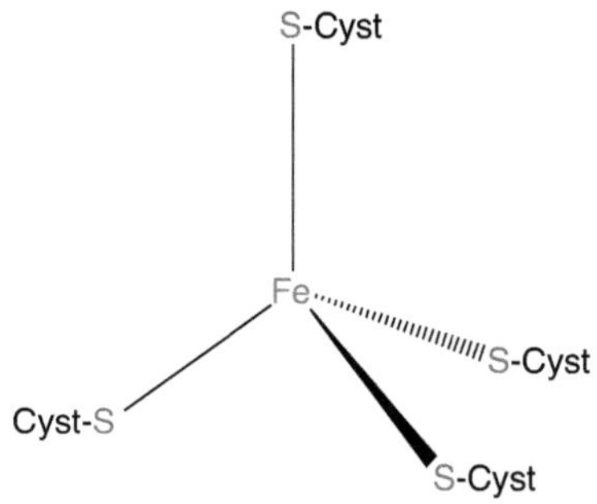

FIGURE 4.20 Representation of rubredoxins iron-sulfur protein [2Fe-2S] proteins.

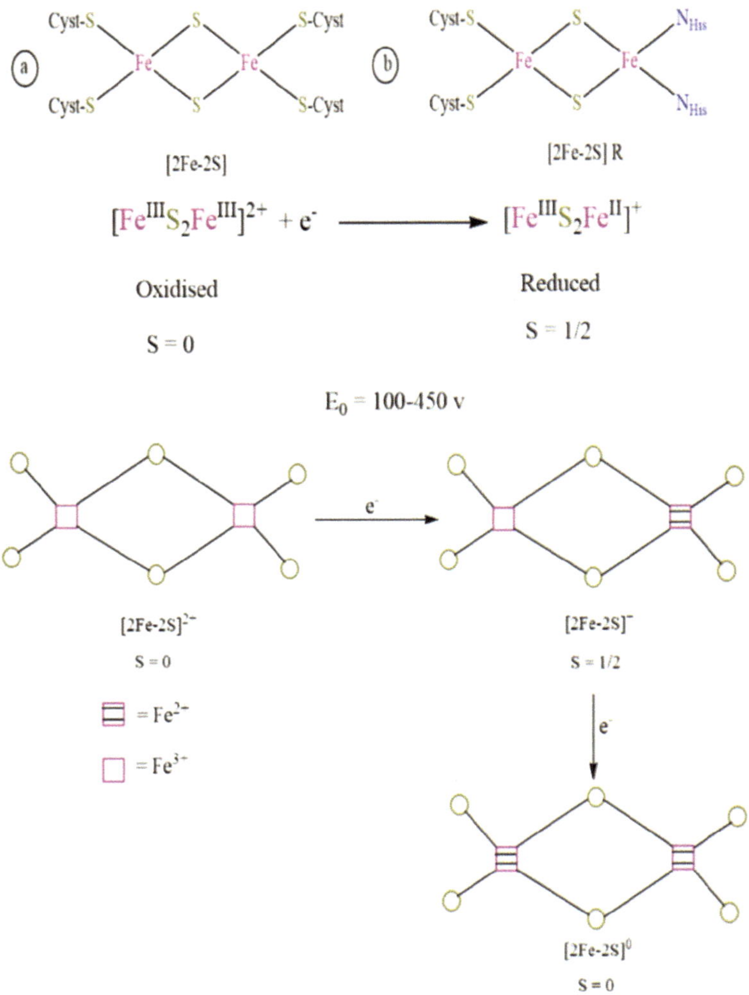

FIGURE 4.21 Representation of ferredoxins iron-sulfur protein.

In general, iron atoms are tetrahedrally ligated with two cysteine groups and two bridged sulfide dianion. The cluster has completely terminal ligation through cysteine groups to each iron. However, there is another subclass, termed Rieske proteins or Rieske centers, which have two histidyl ligands at one Fe site (Figure 4.21). A few cases have been reported in which one non-cysteinyl ligand is attached to the iron metal, such as aspartate in succinate dehydrogenase and arginine in biotin synthetase. [2Fe-2S] proteins act as mediators of electron transfer in processes such as photosynthesis and respiration. The redox potential for a normal [2Fe-2S] cluster (all ligated with cysteine) is in the range of +100 to −460 mV, which is lower compared to Reiske proteins due to the presence of electropositive histidine proteins at the reducible Fe site.

[3Fe-4S] CLUSTERS

These clusters were first reported in the anaerobic nitrogen-fixing bacterium *Azotobacter vinelandii.* In this case, two different subclasses, i.e., cubane and linear types have been recognized in biological systems. Among them, only the cubane type has shown physiological importance. [3Fe-4S] cubane clusters have a defined core of $[Fe_3(\mu_3\text{-}S)(\mu_2\text{-}S)_3]$, which can be compared with the cubane core $Fe_4(\mu\text{-}S)_4$ minus one iron atom, accompanied by cysteine to attain a tetrahedral environment (Figure 4.22). Initially, it was generally believed that the [3Fe-4S] clusters are formed by the removal of one Fe atom from [4Fe-4S] clusters. Structural information elucidates shorter Fe-μ_2S bond lengths (2.24 Å) compared to Fe-μ_3S bond lengths, i.e., 2.29 Å. The average bond length for the terminal Fe-S(Cyst) bond is 2.28 Å, along with two shorter Fe-Fe separations (2.64–2.67 Å) and one longer separation in the range of 2.71 Å. EPR studies show an active oxidized form of the cluster with S = 5/2. Fe^{III} ions are antiferromagnetically coupled to exhibit an S = ½ ground state [17].

[4Fe-4S] FERREDOXIN

[4Fe-4S] proteins are comprehensively involved in electron transfer in various bacteria. In addition to their ubiquity in the function of biological electron transfer, [4Fe-4S] clusters are also involved as catalysts for disulfide reduction and regulatory processes. These proteins have a perfect cubane structure with a $[Fe_4(\mu\text{-}S)_4]$ core. In this case, Fe and S (cysteine) atoms occupy alternate positions of the cube, and every iron atom is ligated with cysteinyl-S protein to adopt a distorted tetrahedral arrangement. The oxidation states of iron observed in these clusters range from +3 to 0. This protein contains two Fe^{2+} and two Fe^{3+} states in its oxidized form, whereas there are three Fe^{2+} and one Fe^{3+} in the reduced state. Active sites of the oxidized form of [4Fe-4S] ferredoxin proteins display EPR-inactive behavior, having a ground state of S = 0. However, the fully reduced form gives an EPR spectrum that matches a ground state spin of ½. [4Fe-4S] ferredoxins undergo one-electron redox cycling between high-potential iron-sulfur proteins $[4Fe\text{-}4S]^{3+,2+}$ (with mid-point potentials ranging from +50 to +500 mv) and $[4Fe\text{-}4S]^{2+}$, pairs (with mid-point potentials in the range of +80 to –710 mV). The magnetic properties of these clusters are less well-known compared to [2Fe-2S] and [3Fe-4S] clusters [17]. The redox interconversion of [4Fe-4S] sites is illustrated below (Figure 4.23).

NITROGEN FIXATION

Biologically, nitrogen is reduced to ammonia under mild conditions (290 k and 0.8 atm) using an enzyme called nitrogenase. Despite an abundant supply of nitrogen on Earth, reduction of nitrogen to incorporate it in the form of nucleic and amino acids cannot be carried out by most organisms. This limitation is mainly due to the high energy requirement (up to 40% of a bacterium's ATP production) for the

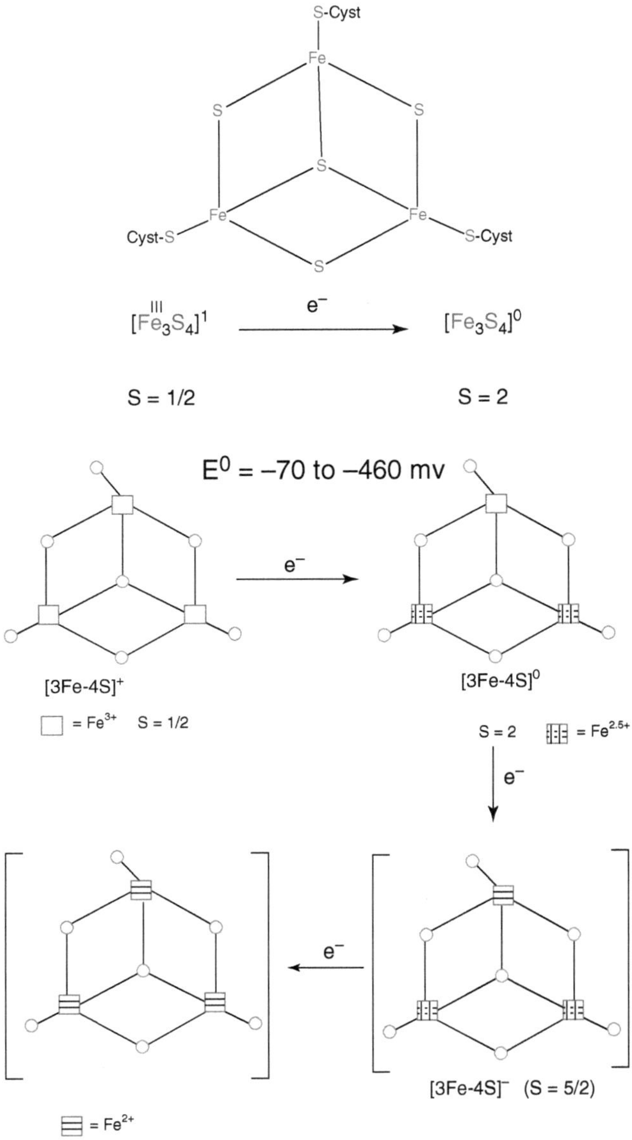

FIGURE 4.22 Representation of [3Fe-4S] iron-sulfur protein.

reduction of nitrogen followed by protonation to yield NH_3 molecules [19–21]. A schematic depiction of nitrogenase enzymes is shown in Figure 4.24.

Ammonia is produced from nitrogen and hydrogen through the Haber-Bosch process in the presence of Fe as a catalyst, using 800 K and 500 atm pressure.

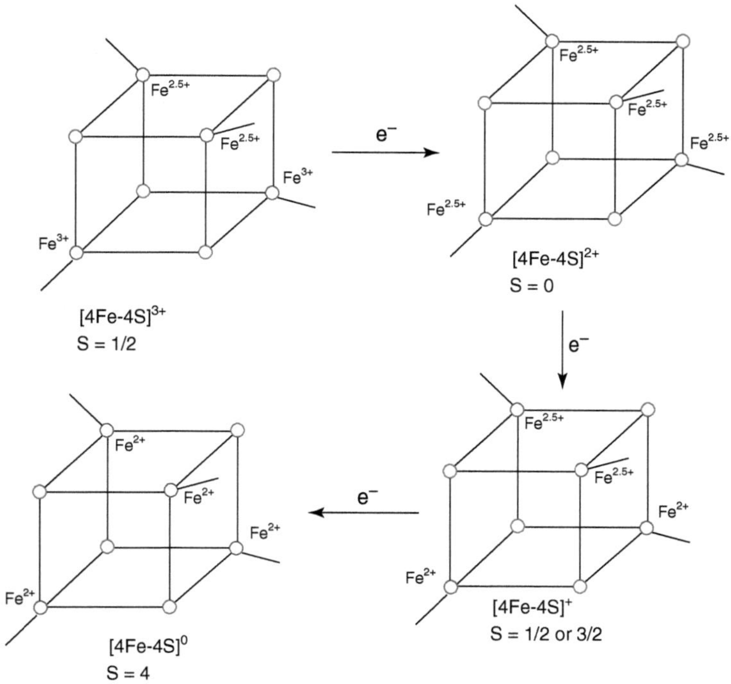

FIGURE 4.23 Representation of [4Fe-4S] iron-sulfur protein.

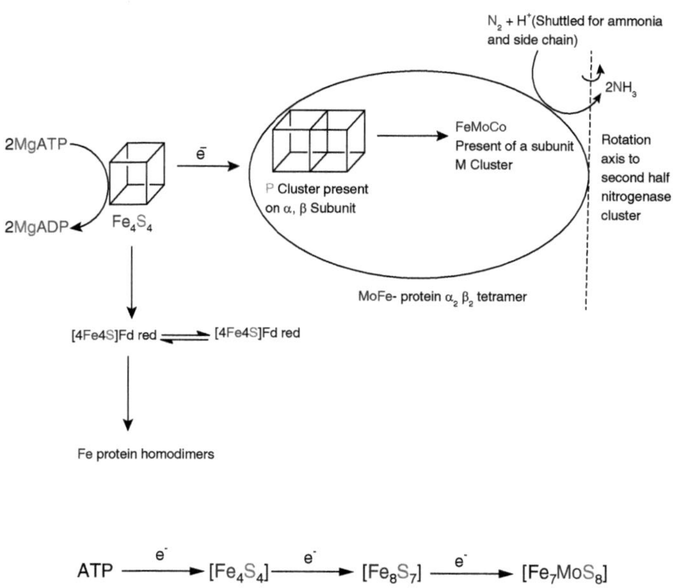

FIGURE 4.24 Schematic representation of nitrogenase protein.

The catalytic reduction of nitrogen molecules to ammonia in biological systems is a thermodynamically favorable process and occur as:

$$N_2 + 8H^+ + 16MgATP \rightarrow 2NH_3 + H_2 + 16MgADP + 16PO_4 \quad (4.6)$$

The above reaction involves the ferredoxin protein for electron transfer as mentioned below:

$$N_2 + 8H^+ + 8Fd_{red} \rightarrow 2NH_3 + H_2 + 8Fd_{ox} \qquad \Delta G = -65.6\,KJ/mol \quad (4.7)$$

Although the biological conversion of nitrogen to ammonia is energetically favorable, the initial step, i.e., the reduction of nitrogen molecules is highly endergonic due to the thermodynamic and kinetic stability of the nitrogen molecule [22].

$$N_2 + H_2 \rightarrow N_2H_2 \; \Delta G = +220\ KJ/mol \quad (4.8)$$

An estimated 175 million metric tons of dinitrogen are fixed annually by nitrogenase enzymes under mild conditions. However, only 50 million metric tons are produced by the Haber-Bosch process at very high temperatures and pressure.

Nitrogenase

The biological fixation of nitrogen was first reported in diazotrophic bacteria and their family. The most studied systems are *Klebsiella pneumoniae* and *Azotobacter vinelandii*. Nitrogenase enzymes present in these bacteria contain either an Fe-Mo or Fe-vanadium core. Mo-containing nitrogenases are composed of two separable proteins: an FeS protein consisting of four irons with a molecular weight of 60 kDa, and a Mo-Fe protein composed of 30 irons and two molybdenum atoms with a molecular weight of 240 kDa. The smaller component, i.e., the iron protein, is γ_2 dimer of identical units. ATP-triggered electron transfer takes place through Fe-S to the M0-Fe cofactor. The second, larger, and more complex part of the nitrogenase subunit—the MoFe protein—transfers electrons first through the P cluster, which contains eight iron and seven sulfur atoms, and subsequently to the M cluster. The latter complex has the stochiometric composition 7Fe:1Mo:8S:1 homocitrate ion. It is suggested that the M-cluster is responsible for binding the substrate, i.e., N_2 and protons are reduced to active sites through amino and side chain hydrogen bonding shuttles [23–26]. The detailed mechanism involved in the reduction and protonation process is as follows:

1. Two moles of Mg-ATP reduce the ferredoxin protein to $[4Fe\text{-}4S]^+$. Conformational changes occur to transfer the electron from the [4Fe-4S] cluster to the next unit.
2. Electron transfer takes place from reduced ferredoxin to MoFe protein. As a consequence, the [4Fe-4S] cluster moves to $[4Fe\text{-}4S]^{2+}$. In short,

two molecules of Mg-ATP transfer one electron, corresponding to the dimeric nature of the Fe protein. Reduced electron potential facilitates electron transfer from the Fe protein to the Mo-Fe protein. Reduction occurs due to the binding of Mg-ATP to the Fe protein, followed by electron transfer to the Mo-Fe protein [22–24] (rate-determining step).

3. In the final stage, electron transfer takes place to dinitrogen substrate at the Mo-Fe proteins, with electron passing through the Mo-Fe protein's P clusters. In this process, the nitrogen molecule is bound to the Fe-Mo cofactor of Mo-Fe protein.

Although the mechanism and sequential changes involved in the reduction of nitrogen and protonation are not well established, it is expected that conformational changes in the Mo-Fe protein are required for electron transfer from the P cluster to the M cluster, coupled with MgATP binding and hydrolysis.

Conclusively, conformational changes in the MoFe protein are required to facilitate electron transfer from the P cluster to the M cluster, with the conversion of ATP to ADP. Electron transfer proceeds from the Fe protein to the electron-buffering capacity of the P and M clusters.

BASIC OF MO-FE PROTEIN

Mo-Fe protein is a tetramer of two α and two β protein subunits. One complete unit is called a component. In this case, two distinct types of reduction centers are present: (i) P Clusters and (ii) Iron-molybdenum cofactors. The P cluster contains two iron-sulfur clusters [22–24]. It acts as a powerhouse to mediate and accumulate electrons to pass them on to the M Cluster. In addition to small molecules of nitrogen, nitrogenase is responsible for a variety of small and unsaturated molecules. Some of these facile conversions are discussed here:

$$C_2H_2 \rightarrow C_2H_4$$

$$N_2O \rightarrow N_2$$

$$HCN \rightarrow CH_3NH_2$$

Iron-sulfur protein acts as an electron mediator coupled between ATP hydrolysis and the Mo Fe protein and also participates in the biosynthesis of the Fe-MoCo protein.

STRUCTURE OF MO-FE PROTEIN

Usually, metal-sulfur clusters are held at the protein surface by coordination through thiolate cysteine residues to the iron center. The basic structure comprises a 4Fe-4S core in a cuboid arrangement with tetrahedral geometry.

TABLE 4.7
Various Oxidation State of 4Fe-4S Clusters

Cluster	Formal Valency	Nature
$[Fe_4S_4]^{3+}$	$3Fe^{3+}$, Fe^{2+}	Oxidized
$[Fe_4S_4]^{2+}$	$2Fe^{3+}$, $2Fe^{2+}$	Intermediate
$[Fe_4S_4]^{+}$	Fe^{3+}, $3Fe^{2+}$	Reduced
$[Fe_4S_4]^{0}$	$4Fe^{2+}$	Super reduction

Two common oxidation states of iron, II and III, result in d^5 and d^6 high-spin configurations. In general, only two oxidation states are stable, but in the case of 4Fe-4S iron-sulfur proteins four states have been identified (Table 4.7). It is worth mentioning that the ferrous state is only stable in protein-bound clusters and not in synthetic models of 4Fe-4S clusters.

P-CLUSTER

The P-cluster of the nitrogenase enzyme is located at the interface of the α and β subunits of the Mo-Fe protein. The P-cluster of the Mo-Fe protein has the formula Fe_8S_7. The basic structure is composed of two fragments of $[Fe_4S_3]$ units bridged by an inorganic sulfur atom. In this case, a [4Fe-3S] cuboid is attached to a [4Fe-4S] cuboid. Three cysteine side chains bind to the iron ions (Figure 4.25a). The cluster exists in P_N and P_o forms (Figure 4.25b), in which the P^N form has only ferrous ions with a total spin of zero (diamagnetic), whereas P_{ox} corresponds to P^{2+} with a total spin of 3 (Table 4.8).

M Center

The M center of the Mo-Fe protein, generally referred to as the Fe-Mo cofactor (FeMoCo), has an $MoFe_7S_9$ core in which the cuboidal cores Fe_4S_3 and $MoFe_3S_3$ are bridged by three sulfides. The M center is quite unusual since it contains three-coordinate iron atoms rather than the expected four-coordination (Figure 4.26). This unsaturation facilitates the attachment of nitrogen for its reduction and protonation processes. The separation between the P and M clusters is 14 Å, making it difficult to explain the mechanistic system of electron transfer. Researchers believe that electron transfer takes place through tunneling between the protein's amino acid side chains. Since two FeMoCo units contain an α, β dimer, the possibility of binuclear coordination of nitrogen to two metal centers, neither linear nor bridged, is feasible [24].

The M center is attached to the protein side chain at two points: Cys 275 (at Fe) and histidine 442 (at Mo). The molybdenum center adopts a distorted octahedral geometry, coordinated with three S^{2-} ligands and chelated homocitrate in addition to His_{447}. The Fe-Fe separation within the cluster is 2.7 Å, whereas across the

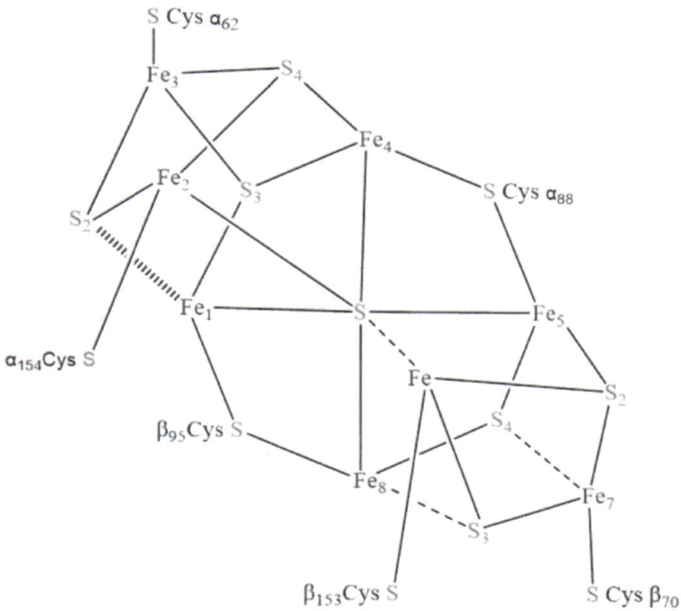

FIGURE 4.25a P^N form of nitrogenase P cluster.

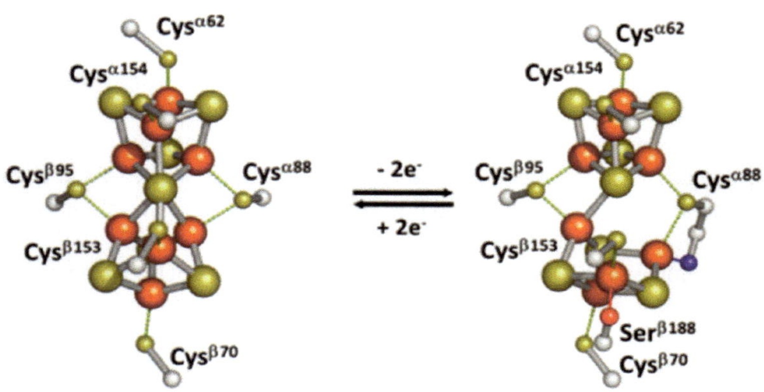

FIGURE 4.25b Schematic diagram of P^N (a) and P^{OX} (b) states of P clusters. (Reused with permission from Ref. [25], Copyright (2023) Elsevier.)

TABLE 4.8
Oxidation and Spin State of P Cluster

State	Oxidation State	Core Oxidation State	Total Spin
P^N (P or P^0)	4 Fe^{2+}	$[Fe_4S_4]^0$ S=0	S_t=0
P^{ox} (P^{2+})	($2Fe^{2+}$, $2Fe^{2+}$)	$[Fe_4S_4]^+$ S=1/2	S_t=3 or 4
	($3Fe^{2+}$, $1Fe^{3+}$)	$[Fe_4S_4]^+$ S=7/2	

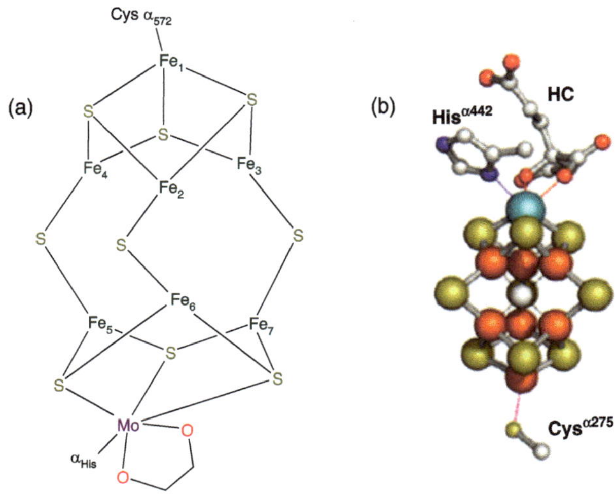

FIGURE 4.26 (a) Schematic diagram of M center (MoFe7S9) of the Fe-Mo cofactor (FeMoCo), (b) Crystal structure of M Centre (Panel b reused with permission from Ref. [25], Copyright (2023) Elsevier.)

cluster, the separation is 3.8 Å. Homocitrate is coordinated to the molybdenum center by hydroxyl and carbonyl oxygen.

Iron Storage

Ferric ion (d^5) is the most environmentally stable form of iron since prebiotic times. The requirement for a large amount of iron used by cells triggered the necessity of iron storage. Hence, nature evolved its own pathway within organisms to store iron in the form of organic chelates. The presence of these species on the surface of water can be related to 'red tide', an explosive 'bloom' of algae (*Gymnodium breve*) that causes mass mortality of fish.

This phenomenon can be correlated with the occurrence of high iron and humic acid in the stream flow. Red tide contains iron-binding siderophores. Similar compounds are present in various aerobic organisms, which solubilize and transport iron (III) in the bloodstream. Depending on their molecular structure and means of chelating iron, these are classified into several groups such as ferrichromes, ferrioxamines, transferrin, etc. (Figure 4.27). These complexing molecules are polydentate ligands with many potential donor atoms for ligating and forming chelates. They readily form stable octahedral complexes with high-spin Fe(III). Polydentate coordination [18] makes these complexes more stable, which is important for performing various biological functions. Their labile nature permits iron to be transported and transferred within bacteria. In higher animals, this function is carried out by transferrin, which is responsible for transporting iron to sites of iron-containing compounds (e.g., hemoglobin and cytochromes) and inserting it via enzymes into porphyrin. Interestingly, iron compounds such as hemoglobin, myoglobin, cytochromes, and ferredoxin employ the 2+ oxidation state of iron, whereas siderophores and transferrin employ Fe(III). The presence of the

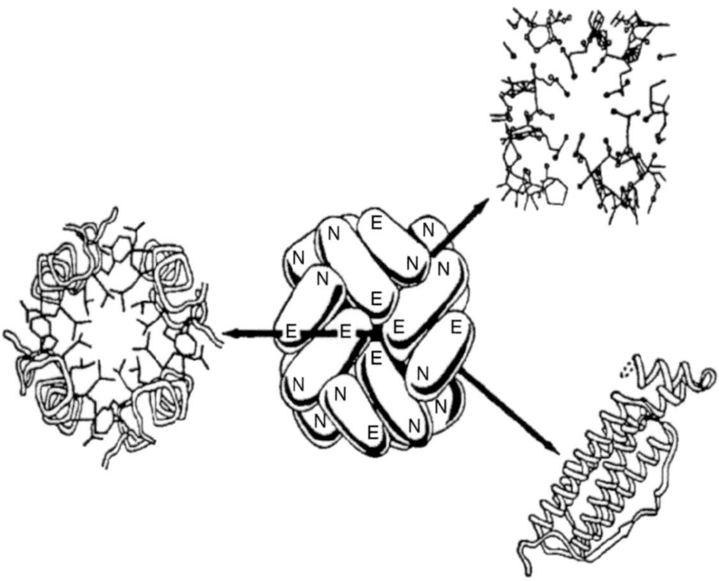

FIGURE 4.27 Structure of apoferritin. (Reused with permission from Ref. [26].)

reduced form of iron within biological systems supports an evolutionary history from the reducing atmosphere of the primitive Earth. However, transferrin and siderophores with Fe(III) advocate for an oxidized external environment. Details about the structure and other physiological parameters are discussed herein.

FERRITIN

In higher classes of animals, iron is stored in the form of an iron complex such as transferrin, which seems to perform the same function as ferrichrome in lower animals [18]. Similarly, higher animals evolved a method of storing iron in the form of ferritin. The storage of iron in protein is known as ferritin. The main requirement of iron is to carry oxygen via red blood cells and circulate it around the body to organs such as the liver, bone marrow, and muscles. A low concentration of iron leads to anemia, in which the body contains very limited red blood cells. Iron deficiency is caused by excessive blood loss or a poor diet. In adverse conditions, the ferritin protein releases the stored iron in a controlled manner. Ferritin is produced by all living organisms, including some archaea, bacteria, algae, higher plants, and animals, but it is found in the spleen and blood in mammals. Ferritin has three compartments: the protein shell, the iron-protein interface, and the ion core. The mineral core (ion core) contains up to 4,500 Fe atoms. In this portion, iron is deposited in the form of $Fe_2O_3(H_2O)_x$ along with phosphate ions. As a sheet, all iron atoms are octahedrally oriented and surrounded by oxygen. The presence of hydroxide and phosphate ions assists in maintaining the charge and binding at the protein surface.

The protein sheath is hollow and globular in nature, having a mass of about 474 kDa with 24 subunits. Its internal diameter and external diameter are 8 and 12 nm, respectively. In vertebrates, two subunits are observed: light (L) and heavy (H), with masses of 19 and 21 kDa, respectively. The structural features of the two cylindrical subunits are different. One end is represented as N, which refers to the nonpolar N-terminal end of the protein, and the other end is designated as E, referring to the nonpolar helical segment. The present subunits link together in a circular manner to form a hollow sphere, which contains two-to-four-fold symmetry axes. All 24 subunits are arranged in such a manner that three subunits meet with their N-ends to form a polar channel through which iron can pass in or out. There are also six nonpolar channels formed by four subunits with their E-ends, creating nonpolar channels, whose roles are not yet established [24–26]. The proposed mechanism of Fe in ferritin involves the transport of iron in and out as Fe(II), which is soluble at neutral pH. The ferroxidase centers are the places where the oxidation of Fe(II) to Fe(III) takes place, and they are present in each subunit. The oxidation of Fe(II) to Fe(III) occurs through inner-sphere electron transfer and coordination with O_2.

$$2\mathrm{Fe(II)} + 2\mathrm{H}^+ \xrightarrow[\mathrm{O_2}]{} 2\mathrm{Fe(III)} + \mathrm{H_2O_2}$$

Oxidation of Fe(II) to Fe(III)

Ferritin, as the name suggests, contains ferric ions and its best-established physiological function is to store iron. It has two roles: it supplies an elemental reserve that can be used to synthesize molecules like hemoglobin, cytochromes, and iron-sulfur complexes as needed. Additionally, it offers a way to protect the cell from potentially harmful material. Iron within ferritin is relatively inert. Although there is no clear amino acid sequence relationship between ferritin and bacterioferritin, even bacteria contain ferritin-like molecules, which seem to have evolved rather separately. Interestingly, iron in a second storage form called hemosiderin, which appears to be a byproduct of the breakdown of ferritin and is contained within secondary lysosomes or siderosomes, is claimed to be less accessible than that of ferritin. If this is true, the rise in the ratio of hemosiderin to ferritin with higher iron loading may lead to a secondary protective mechanism. It is interesting to note that hemosiderin implies its origin in hemoglobin, not iron.

Functions

- All cells contain ferritin, and its function is to store iron in a nontoxic form and transfer it to all areas where it is required. Free iron acts as a catalyst in the formation of free radicals via the Fenton reaction, and that free iron is also very toxic to the cell. Vertebrates have various sets of protective mechanisms to bind iron in tissue compartments. Hence, iron is stored in protein complexes as ferritin or hemosiderin inside the cell. Apoferritin stores ferric ions after binding to ferrous ions inside the cell.

- The ferritin concentration increases in the presence of infection or cancer. The concentration of ferritin is increased due to the endotoxin, which is a gene coding for ferritin. The endotoxin present in the organism Pseudomonas causes a significant drop in plasma ferritin levels within 48 h of infection. As a result, the infected body denies the infective agent, hindering its metabolism.
- Under stress conditions such as anoxia, it has been observed that the concentration of ferritin increases continuously, implying that it is an acute phase-protein.
- In gastropods such as snails it is observed that the protein compartment in the egg yolk is ferritin. It is produced inside the midgut gland, is secreted into the hemolymph, and travels to the eggs.

TRANSFERRIN

Iron is an important element in several metabolic pathways. Iron balance is very important because any change, whether too little or too much, can harm the body. Transferrin is very close to ferric iron, so there is not much free iron in the body. This is because transferrin binds almost all plasma iron. Transferrin is a glycoprotein in which iron is present in the Fe(III) state. Examples include serum transferrin (blood plasma), ov0transferrin (egg white), and lactoferrin (milk). All these are glycoproteins with a molar mass of approximately 80 kg/mol, containing two separate and equivalent binding sites for Fe ions. During complexation, Fe(III) ions at each site bind with HCO_3^{2-} or CO_3^{2-} and release H^+ ion.

$$\text{R-TF} + \text{Fe(III)} + HCO_3^{2-} \rightarrow \text{TF} - \text{Fe(III)} - CO_3^{2-} + H^+$$

TF-Transferrin

Transferrin has two parts, i.e., the N-lobe and C-lobe proteins, which are formed due to gene duplication because the structure of the 1st half of the molecule overlaps with the second half. Two domains of each half, 1 and 2, together form a cleft with binding sites for iron ions (Figure 4.28).

The active sites of Fe(III) atoms are coordinated through the side chains of amino acids from both domains and the connecting portion. The geometric

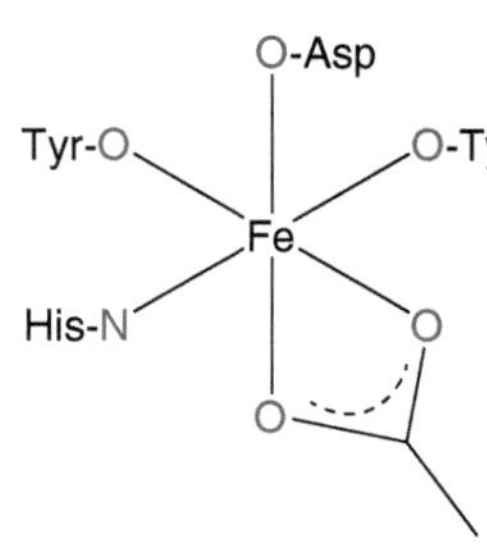

FIGURE 4.28 Structure of transferrin.

conformation observed includes the presence of the ligands aspartate (carboxylate), two tyrosines (phenolate-O), and histidine (N-His). Only one aspartate atom is coordinated through an oxygen atom. The different protein ligands such as aspartate and two tyrosine create a surrounding environment where Fe(III) obtains a distorted octahedral geometry. The coordination is completed by carbonate, which acts as a synergistic ligand, and Fe binding depends on its presence. In certain cases, phosphate bonds instead of carbonate.

Release Mechanism

The release of iron-bound transferrin begins with the formation of clathrin-coated pockets and the vesicle's internalization into the cytoplasm. Due to a decrease in pH, the clathrin coating of the vesicle is lost. The reduction of pH by hydrogen ion proton pumps (H+ ATPase) to 5.5 results in the discharge of iron from iron-bound transferrin vesicles. In addition, transferrin's affinity for iron is diminished when it binds to transferrin receptors. The stability of the chelates is highly pH-dependent. At pH 7, the Fe^{3+} chelate is very stable, but at lower pH (5.5), Fe dissociates from the chelate. Hence, at lower pH, bound CO_3^{2-}/HCO_3^{2-} undergoes acid-catalyzed dissociation. Once endocytosed, two pathways are possible: degradation or recycling. Early and late endosomes dissociate transferrin from ferric ions along the degradation pathway. Now, iron can be used as a storage material or incorporated into hemoglobin. The recycling pathway entails transferrin recycling. After iron dissociation, transferrin is referred to as apotransferrin. Due to its high affinity for its receptors at a reduced pH, apotransferrin remains bound to its receptor. It returns to the plasma membrane while still attached to its receptor. At a neutral pH, apotransferrin dissociates from its receptor to enter the circulatory system, replenish iron, and continue the cycle [10,18]. All transferrin receptors eventually follow the receptor degradation pathway. In the bone marrow, an erythroid precursor is an example of such a cell.

Functions

- At neutral pH, Fe(III) ions are insoluble. They are soluble only when iron binds to transferrin
- Transport of iron to all biological tissues through absorption, utilization and storage.
- Avoidance of reactive oxygen species formation.
- Chelation of free toxic iron, functioning as a scavenger.
- Transferrin is a component of the innate immune system, and its binding to iron inhibits bacterial survival.
- Transferrin is an indicator of inflammation, and its concentration decreases during inflammation.

SIDEROPHORES

The transport of iron in microorganisms such as bacteria, fungi, etc., takes place with the help of a chelating agent called siderophores. The chelating agents hydroxamates and substituted catechols are produced by bacteria, yeast, and fungi [1,10,18]. The production of these complexing ligands increases when the concentration of Fe is low in the organism. The siderophores responsible for promoting growth are called sideramines. The antibiotics sideromycins (Fe-free ligands) function by strongly binding to the ligand, making Fe unavailable for bacterial development. The siderophores are also known as siderochromes due to their intense color (red-brown).

These complexing ligands bind to Fe(III) ions to form a very stable complex. The formation constants for Fe(II) complexation with the ligand are much lower, so the reduction of Fe provides a route for its release. The ligands capture the Fe^{3+} ion at the cell membrane and transport it into the cell. The charge type of the complex determines the process of transmission through the cell membrane. The overproduction of the ligand in the presence of low concentrations of Fe(III) results in their release into the surrounding media to dissolve Fe(III).

The hydroxamate complex is called ferrichrome, in which hydroxamide groups bind to the peptide ring through side chains. Ferrichrome differs in ring substitutes and has the absolute configuration in which the chelate rings surround Fe(III) ions (Figure 4.29). The hydroxamates are part of the peptide chain of ferrioxamines. Less stable hydroxamate complexes are formed during the reduction of Fe(III) to Fe(II) ions, in which iron is released from the complex, and the ligands are available for reuse. The Fe released from the complex is the result of the ligand being destroyed by ester hydrolysis by specific enzymes. The hydrolysis products are not employed in the production of the ligand.

Biological Function

- A sufficient amount of iron is required for plant growth in soil. Iron deficiency in plants is a problem in calcareous soil due to the presence of less soluble iron hydroxide. In such conditions siderophores are secreted by graminaceous plants in the soil. During the growth of graminaceous plants in iron-deficient soil, the siderophore is secreted into the rhizosphere through the roots. Upon finding a sufficient amount of iron, the iron(III)-siderophore complex transports iron across the cytoplasmic membrane. Then, iron(III) is reduced to iron(II) and transferred to nicotianamine. In the last step, nicotianamine transfers iron to all parts of the plant through the phloem.
- Iron is not easily available in the environment, and it is an important nutrient for Pseudomonas aeruginosa bacteria. In such situations. Pseudomonas aeruginosa produces siderophores for binding and transporting iron, but it is not necessarily the case that the produced siderophores directly benefit from the entire iron supply. The energy is expanded to all cells of the bacterium, which also require siderophores.

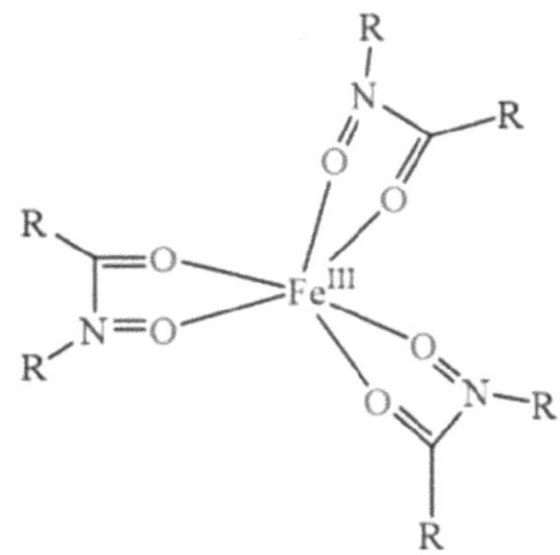

Iron (III) Hydroxamate Complex

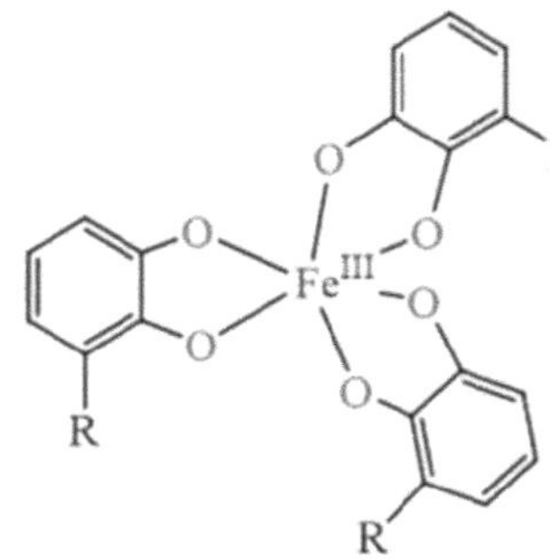

Iron (III) Catechol Complex

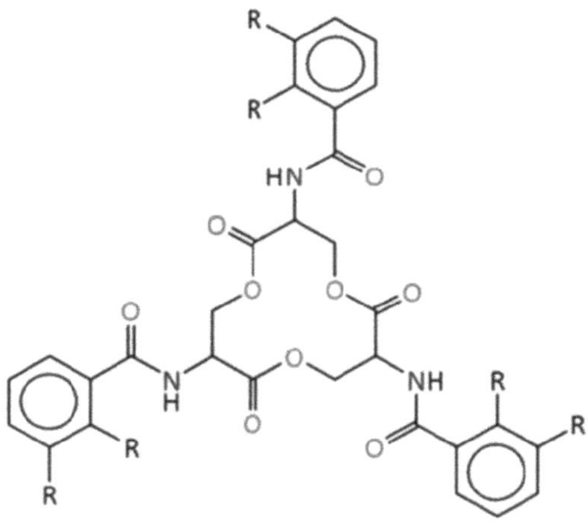

(a) R = OH

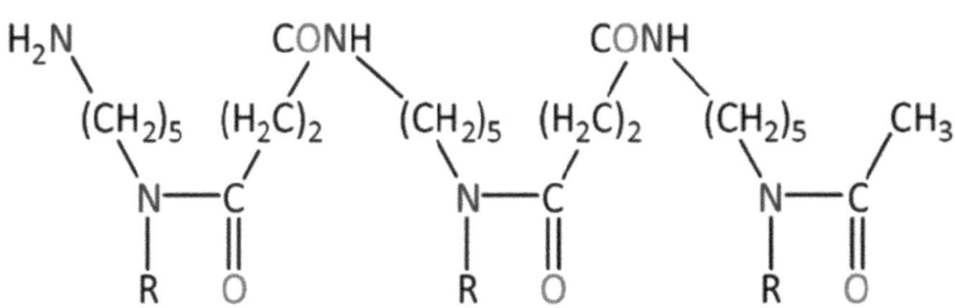

(b) R = OH

FIGURE 4.29 Structure of siderophores (a) desferrioxamine (b) enterobactin.

- Genes involved in the production and uptake of microbial siderophores are suppressed due to the limitation of iron in the environment. To overcome this problem, the production of siderophores and uptake proteins takes place. Iron is released due to Fe^{III} reduction to Fe^{II} in the cytoplasm of the cell, especially in the case of hydroxamates and carboxylates. Similarly, in the case of catechols such as ferric enterobactin, the reduction potential is low for flavin adenine dinucleotide reducing agents hence, iron is released due to enzymatic degradation.

PEROXIDASE

Another important class of reduced oxygen is hydrogen peroxide. Several enzymes have been studied that either destroy the peroxide to oxygen and water or use it as an oxidant (Table 4.9). Here, we are discussing only the enzymes that contain a heme group but differ in their axial nature (Figure 4.30).

$$2H_2O_2 \rightarrow 2H_2O + O_2 \quad \text{Catalase}$$

$$AH_2 + H_2O_2 \rightarrow A + 2H_2O \quad \text{Peroxidase}$$

$$RCOOH + 2H^+ + 2e^- \rightarrow ROH + H_2O \quad \text{Cytrochrome}$$

TABLE 4.9
Details of Various Heme Containing Iron-Based Enzymes

Enzyme	Molecular Mass	Heme/ Non-Heme	No. of Fe Atoms per Molecules	Functions
Cytochrome C peroxidase	>100	Heme	02	Terminal oxidation
Horse radish peroxidase	–	–	–	Metabolism of H_2O_2
Catalase	260	Heme	01	Metabolism of H_2O_2
Cytochrome C	12.5	Heme	01	Electron transfer

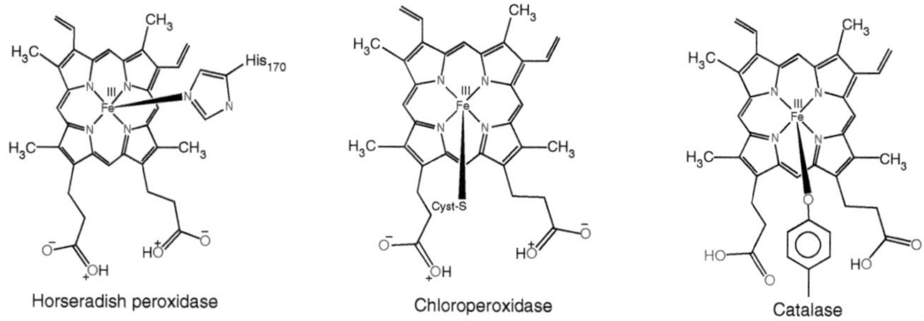

FIGURE 4.30 Structures of various peroxidases.

The structure of the peroxidase enzyme is not well known, but spectroscopic investigations show that the axial position is occupied by histidine, similar to hemoglobin [25,26]. However, the opposite axial position remains vacant in resting Fe(III). The vacant position is used to provide a binding site for H_2O_2 during the catalytic activity of the enzyme.

Mechanism

1. The First step involves the two-electron oxidation of the enzyme [represented as E Fe(III)] by hydrogen peroxide with the loss of one molecule of water.
2. The nature of the electron-oxidized form, E Fe(IV)=0, is quite interesting. The iron-oxo moiety is responsible for one of the two electrons and the other comes from the porphyrin ring to yield a cation radical. This state is called compound I.
3. The two-electron-oxidized species, compound I, undergoes several reactions with a variety of reacting species, e.g., XH_2. When reacting with compound I, it results in the production of A and $2H_+$ with the subsequent reaction ending in H_2O and E Fe(III) (Figure 4.31).

In the case of peroxide enzymes, these reaction species are alkylamines or sulfides, whereas chloro peroxidase enzymes catalyze the conversion of chloride to hypochlorite, which, when transferring the halogen to the oxygen substrate, produces hydroxide.

CYTOCHROME P-450

Cytochrome P-450 is another class of enzyme that contains heme as a cofactor at the active site. The primary function of this enzyme is to catalyze the oxidation of substrates (Figure 4.32). In mammals, it is important for the oxidation of steroids, fatty acids, and the clearance of xenobiotics through the body [15].

$$\text{E Fe(III)} + H_2O_2 \rightleftharpoons \text{E P Fe}^{IV}=O + 2H_2O$$

Compound - 1

$$\text{Compound - 1} + AH_2 \rightleftharpoons A + H_2O + \text{E (Fe(III))}$$

$$\text{Compound - 1} + H_2O_2 \rightleftharpoons \text{E Fe}^{III} + H_2O + O_2$$

FIGURE 4.31 Steps involved in H_2O_2 metabolism.

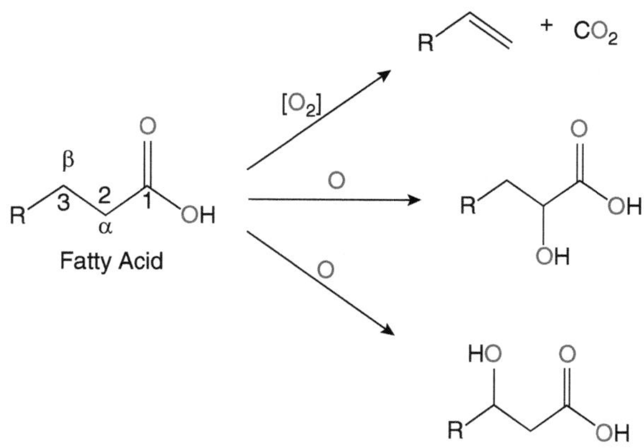

$$RH + O_2 + NADPH + H^+ \longrightarrow ROH + H_2O + NADP^+$$

FIGURE 4.32 Various biological reactions catalyzed by cytochrome P450.

STRUCTURE

The folded structure around the heme cofactor and iron atom protects the active site from the surrounding solution with low permittivity. Iron is octahedrally oriented, with the equatorial plane similar to the hemoglobin structure; it is coordinated to the cysteine thiolate ligand system, which acts as a bridge to tether the iron atom from the sheathed protein (Figure 4.33). The distal axial position is open to coordinate with oxygen. The incoming substrate, present in a hydrophobic pocket, is positioned nearly 5 Å from the iron site.

MECHANISM

1. The substrate is available in the hydrophobic pocket in close proximity to the heme group, opposite the axial cysteine thiolate. Substrate binding triggers conformational changes in the active site that consequently displace the axial water molecule and change the configuration of the iron atom from Fe(III) low spin to high spin Fe(II).
2. Substrate binding initiates electron transfer from NADPH through cytochrome P-450 family enzymes.
3. Molecular oxygen coordinates to ferrous iron Fe(II), resulting in an oxygenated adduct.
4. A second electron is transferred, from either cyto P-450 or ferredoxin, reducing the dioxygen adduct to a peroxo moiety.

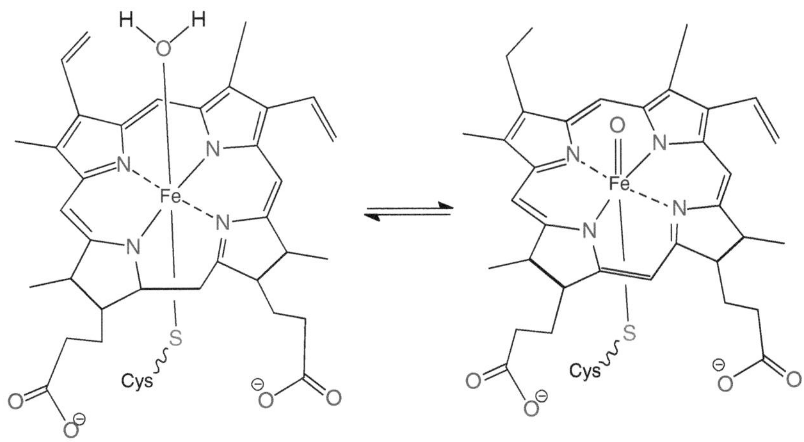

FIGURE 4.33 Structure of cytochrome P-450.

5. The highly reactive peroxo form gets protonated, releasing one water molecule and resulting in another active species (radical) called P-450 compound 1. The latter compound exhibits Fe(IV) behavior (Figure 4.34).
6. The Fe(IV) species binds to the substrate to inert oxygen with the loss of R-OH. After releasing the hydroxylated product from the active site, the enzyme returns to its original form by coordinating a water molecule at the vacated position, bringing the catalytic cycle back to its original resting state [26].

CONCLUSIONS

Three different domains of science, i.e., chemistry, biology, and biochemistry, have a common interest in exploring the various roles of iron in biological systems. In the current chapter, the qualitative aspects of various iron-based enzymes, iron proteins involved in electron transfer, and their storage have been discussed. The primary focus is related to hemoglobin and myoglobin, their structure, functioning, and factors affecting the binding of oxygen. Hemoglobin is responsible for carrying oxygen from the bloodstream to the body, and myoglobin governs the storage of oxygen in muscle tissue. Depending on the cooperativity of oxygen, an explanation related to the preferential binding of heme over CO has been provided. Two probable explanations for this ubiquitous behavior are:

1. The presence of hydrogen bonding and electrostatic forces of attraction
2. Bending coordination with steric considerations that disfavor CO coordination

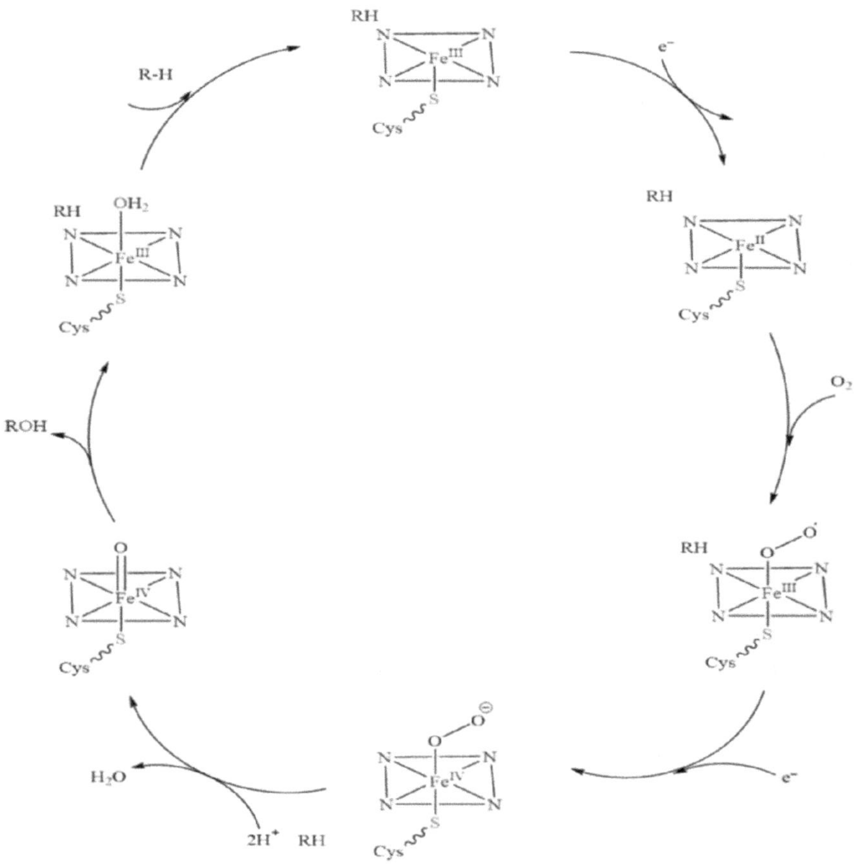

FIGURE 4.34 Catalytic cycle for cytochrome P-450.

REFERENCES

1. R. Crichton, *Iron Metabolism: From Molecular Mechanisms to Clinical Consequences*, John Wiley & Sons, Ltd, United States, 2009, Pages 1–461.
2. S. J. Lippard, J. M. Berg, *Principles of Bioinorganic Chemistry*, University Science Books, Mill Valley, CA, 1994, Pages 1–585.
3. R. M. Roat-Malone, *Bioinorganic Chemistry: A Short Course*, John Wiley & Sons, Inc., Hoboken, NJ, 2002, Pages 1–365.
4. W. Kaim, B. Schwederski, A. Klein, *Bioinorganic Chemistry: Inorganic Elements in the Chemistry of Life*, John Wiley & Sons, Inc., Hoboken, NJ, 2013, Pages 1–426.
5. G. Palmer, J. Reedijk, Nonmenclature of electron-transfer proteins: Recommendations 1989, *Eur. J. Biochem.* 1991, 200, 599–611.
6. J. P. Collman, R. Boulatov, C. J. Sunderland, L. Fu, Functional analogues of cytochrome c oxidase, myoglobin, and hemoglobin, *Chem. Rev.* 2004, 104, 561–588.
7. G. L. Klippenstein, Structural aspects of hemerythrin and myohemerythrin, *Am. Zool.* 1980, 20, 39–51.

8. I. M. Klotz, G. L. Klippenstein, W. A. Hendrickson, Hemerythrin: Alternative oxygen carrier: Nature has developed an effective transport protein with a binuclear iron center in place of a heme, *Science* 1976, 192, 335–344.
9. R. E. Stenkamp, Dioxygen and hemerythrin, *Chem. Rev.* 1994, 94, 715–726.
10. Voet D., J. G. Voet, Biochemistry, John Wiley & Sons, New York 1990, chapter 9.
11. A. Ortiz, Derivation of Hill's equation from scale invariance, *J. Uncertain Syst.* 2013, 7, 198–202.
12. M. K.-S. Leow, Configuration of the hemoglobin oxygen dissociation curve demystified: a basic mathematical proof for medical and biological sciences undergraduates, *Adv. Physiol. Educ.* 2007, 31, 198–201.
13. H. A. Saroff, Action of hemoglobin. Cooperative and Bohr effects, *J. Phys. Chem.* 1972, 76, 1597–1607.
14. P. D. Josephy, *Biochem. Edu.* 1992, 20, 91–93.
15. P. R. Ortiz de Montellano, *Cytochrome P450: Structure, Mechanism, and Biochemistry* (Third Edition), Kluwer Academic/Plenum Publishers, New York, 2005.
16. T. D. Porter, M. H. Coon, Cytochrome P-450. Multiplicity of isoforms, substrates, and catalytic and regulatory mechanisms, *J. Biol. Chem.* 1991, 266, 13469–13472.
17. M. K. Johnson, A. D. Smith, Iron-Sulfur proteins, *Encycl. Inorg. Bioinorg. Chem.* 2011, 1–31.
18. J. E. Huheey, E. A. Keiter, R. L. Keiter, *Principles of Structure and Reactivity, Inorganic Chemistry* (Fourth Edition), 2005, Pages 1–889.
19. R. R. Crichton, *Biological Inorganic Chemistry*, Elsevier, Oxford, 2008, Pages 211–240.
20. R. J. Ouellette, J. D. Rawn, *Principles of Organic Chemistry*, Elsevier, Amsterdam, Netherlands, 2015, Pages 371–396.
21. M. Momenteau, C. A. Reed, Synthetic heme-dioxygen complexes, *Chem. Rev.*, 1994, 94, 659–698.
22. D. M. Kurtz, *Dioxygen-binding Proteins, Comprehensive Coordination Chemistry II*, Pergamon, Oxford, UK, 2003, 8, Pages 229–260.
23. M. T. Wilson, B. J. Reeder, *Encyclopedia of Respiratory Medicine*, Academic Press, Cambridge, 2006, Pages 73–76.
24. K. Min, D. Yeo, J. K. Yoo, B. D. Johnson, C. H. Kim, Would a right shift of the oxy-hemoglobin dissociation curve improve exercise capacity in patients with heart failure? *Med. Hypotheses*, 2020, 134, 109423.
25. Y. Hu, M. W. Ribbe, Nitrogenase assembly, *Biochim. Biopsy. Acta Bioenerg.*, 2013, 1827(8–9), 1112–1122.
26. P. M. Harrison, A. Treffry, T. H. Lilley, Ferritin as an iron-storage protein: mechanisms of iron uptake, *J. Inorg. Biochem.* 1986, 27, 287–293.

5 Role of Cobalt in Biology

INTRODUCTION

Cobalt, a transition element positioned centrally in the periodic table, plays a crucial role in diverse biological systems. Its intriguing chemical properties and versatile coordination behavior position it as indispensable for several biological processes. Its most common oxidation states are 2+ and 3+; however, it can exhibit oxidation states from 1– to 4+. Cobalt, along with nickel, has been found to play a crucial role in the metabolism of methane, carbon monoxide, and hydrogen, which were particularly abundant during the pre-oxygen era. The involvement of cobalt in these metabolic processes highlights its importance in anaerobic microorganisms [1]. For instance, methanogenic archaea employ cobalt- and nickel-containing metalloenzymes to catalyze the conversion of carbon dioxide and hydrogen into methane. Certain bacteria feature nickel- and cobalt-containing carbon monoxide dehydrogenase metalloenzymes, which catalyzes the oxidation of carbon monoxide to carbon dioxide, providing necessary energy for microbial growth [2]. Though anaerobic bacteria exhibit high levels of cobalt, higher organisms such as mammals have significantly lower cobalt levels compared to other essential elements like zinc, iron, and copper. This difference in cobalt levels between anaerobic bacteria and mammals may be attributed to the fact that, in the post-photosynthetic era, many of cobalt's functions, such as acid-based chemistry, could be fulfilled by zinc, while redox functions can be performed by iron or copper. Despite the fact that cobalt is less significantly encountered in metalloenzymes, it is crucial for the synthesis and function of a number of metalloenzymes, including cobalamin. In this chapter, we will discuss some important aspects of cobalt-containing metalloenzymes.

DISTRIBUTION AND BIOAVAILABILITY OF COBALT IN BIOLOGICAL SYSTEMS

The presence of cobalt-containing compounds can be observed in various biological systems, including microorganisms, plants, and animals. The bioavailability of cobalt is significantly influenced by factors such as its chemical form, dietary sources, and interactions with other nutrients. Whereas inorganic forms of cobalt salts, such as cobalt chloride or cobalt sulfate, can be readily absorbed through the gastrointestinal tract, the absorption of organic cobalt compounds, such as cobalamin, is more difficult and requires specific binding proteins and receptors for absorption. Dietary sources play a crucial role in cobalt bioavailability.

DOI: 10.1201/9781003432333-5

Animal-derived foods, including meat, fish, and dairy products, are generally considered more bioavailable sources of cobalt due to their higher cobalamin content. Plant-based sources, such as legumes and leafy greens, may contain cobalt but often in lower amounts and in less bioavailable forms. Furthermore, it is also observed that the presence of vitamin C and dietary proteins can enhance cobalt absorption, while high levels of iron or zinc may reduce cobalt absorption. Besides this, factors such as age, health conditions, and genetic variation can also affect the efficiency of cobalt absorption. For instance, individuals with pernicious anemia, a condition characterized by impaired cobalamin absorption, may experience reduced cobalt bioavailability. Overall, the bioavailability of cobalt is a complex phenomenon influenced by various factors and understanding these factors is essential for assessing its dietary requirements and potential health implications [3].

HISTORICAL BACKGROUND OF COBALT AS BIOLOGICALLY IMPORTANT METAL

The significance of cobalt in biology was first recognized in the early 20th century when researchers investigated the causes of pernicious anemia which was an invariably fatal disease at that time. In their investigations, they discovered that the disease could be treated by supplementing the diet with liver. Around 1925, scientists identified a compound present in liver, which was later named cobalamin or vitamin B12, as the crucial factor responsible for the therapeutic effects against pernicious anemia [4]. It was also realized that cobalamin performs a crucial role in red blood cell synthesis. The purification and isolation of vitamin B12 were later achieved at Merck laboratories, marking significant milestones in understanding cobalt's role in biology. However, unraveling the precise structure of vitamin B12 proved to be a gigantic task. It was the courageous and brilliant work of Dorothy Hodgkin and her colleagues that allowed the determination of the crystal structure of vitamin B12 using the single-crystal X-ray diffraction technique [5].

Determining the crystal structure of a molecule as complex as vitamin B12 was an extraordinary feat, requiring immense dedication and ingenuity. At the time, crystal structure determinations were far from routine, and both X-ray and computer equipment were still in the early stages of development. The effort to synthesize vitamin B12 was equally arduous. It took more than a decade and involved over 90 separate reactions performed by a team of over 100 researchers, including the eminent chemist R. B. Woodward at Harvard [6].

The discovery of vitamin B12 as an anti-pernicious anemia factor, the elucidation of its structure, the synthetic efforts to reproduce it, and the subsequent quantum mechanical studies on vitamin B12 were recognized with four Nobel Prizes—one in medicine and three in chemistry, as listed in Table 5.1. The breakthroughs in understanding the structure and synthesis of vitamin B12 inspired further research in the field, driving advancements in both biological and synthetic chemistry.

TABLE 5.1
Noble Prizes Related to Coenzyme B12

Contributors	Year	Field	Contributions
Whipple, Minot, and Murphy	1934	Physiology and medicine	The discovery of 'anti pernicious anemia factor'
Dorothy Crowfoot Hodgkin	1964	Chemistry	Structural determination of important biochemical compounds
R. B. Woodward	1965	Chemistry	Art of organic synthesis
K. Fukui and R. Hoffman	1981	Chemistry	Quantum mechanical study of chemical reactivity

These pivotal discoveries laid the foundation for unraveling cobalt's essential role in enzymatic reactions and metabolic pathways, further expanding our understanding of its vital biological importance across various organisms.

STRUCTURAL FEATURES OF COBALAMIN

Cobalamin plays a crucial role in the metabolism of all kinds of organisms. While microorganisms are the primary natural sources of cobalamin derivatives, these derivatives function as vitamins for other B12-dependent organisms [7]. Cobalamin-dependent enzymes are responsible for catalyzing various reactions, which are important in almost all organisms, including humans. Therefore, understanding the structure and chemistry of cobalamin is of prime importance.

Cobalamin is a complex as well as structurally intriguing molecule that contains a central cobalt ion (Co). The core structure of cobalamin consists of a modified tetrapyrrolic macrocycle known as the corrin ring. This corrin ring contains four pyrrole rings connected by methylene bridges, creating a planar structure. The tetrapyrrole ring found in vitamin B12 exhibits an intriguing characteristic. It has undergone a ring contraction process, which contrasts with the tetrapyrrole system observed in heme and chlorophylls. In the corrin ring, one of the bridging carbons that typically connects the four pyrrole rings has been eliminated. As a consequence, the resulting tetrapyrrole ring in vitamin B12 is not only contracted but also lop-sided, providing it with distinct structural features. The central cobalt ion resides within the corrin ring, coordinated by four nitrogen atoms derived from the pyrrole rings. The fifth coordination to Co comes from a nitrogen atom from a 5,6-dimethylbenzimidazole nucleotide (Dmb) covalently linked to the propionate side chain of the corrin ring through an aminopropanol linker. Incorporating this nucleotide is essential for the stability and functionality of cobalamin. The sixth coordination in cobalamin varies depending on the specific form of the molecule. It is cyanide (CN) in the case of vitamin B12, whereas it is 5′-deoxyadenosine ligand in the case of adenosylcobalamin. The overall structure of adenosylcobalamin is represented in Figure 5.1. In other forms such as methylcobalamin, methyl group occupies the sixth position. Some bacteria develop other variations

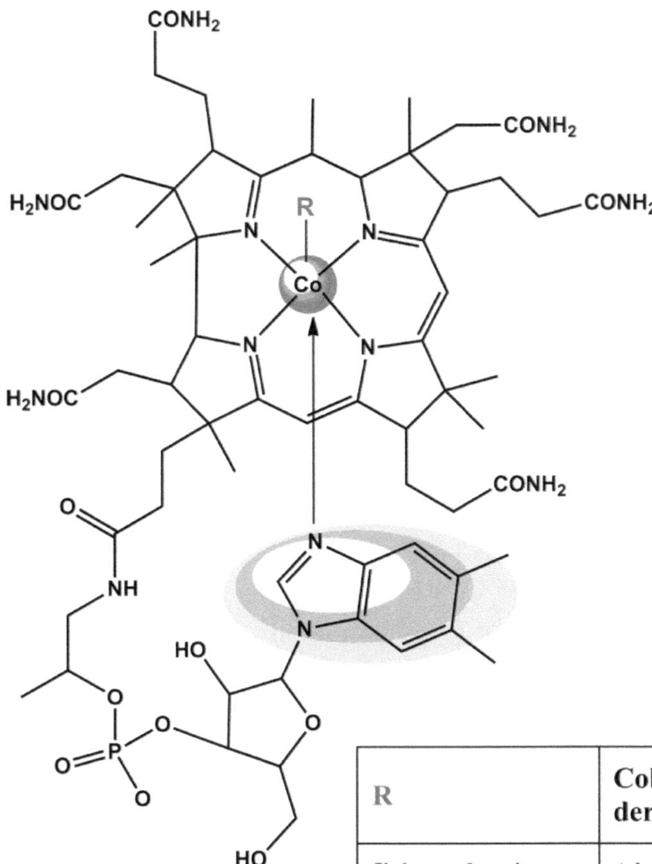

R	Cobalamin derivative
5'-deoxyadenosine	Adenosylcobalamin
CN	cyanocobalamin
CH_3	Methylcobalamin
OH	Hydroxycobalamin

FIGURE 5.1 Structure of vitamin B12.

in cobalamin by substituting dimethylbenzimidazole base with other bases such as purines/phenol systems, etc. [4]. The free cobalamin can exist either in the base-on or base-off configuration. The base-on configuration refers to the coordinated state of the dimethylbenzimidazole base, while the base-off configuration is the uncoordinated state of the dimethylbenzimidazole base [8]. The transition between the base-on and base-off configurations is important for the transportation, reactivity and functionality of vitamin B12. In mammals, cobalamin transport proteins specifically bind and identify the base-on form of cobalamin. On the other hand, certain cobalamin-dependent enzymes, like methyltransferases, bind their organometallic cobalamin cofactors in a base-off configuration. Notably, switching from the base-on to the base-off configuration increases the surface area by approximately 300 $Å^2$. This is particularly beneficial in relieving the geometric strain on the nucleotide loop present in the base-on form [7].

Another notable feature of cobalamin is the presence of a carbon-metal (Co) bond at the sixth position. Such metal-carbon bonds are quite an uncommon feature in biological systems, adding to the intriguing structural features of cobalamin. This coordination also constitutes the primary active biological forms of cobalamin [7]. For instance, methylcobalamin serves as a vital cofactor in numerous methyltransferase reactions, whereas adenosylcobalamin acts as a coenzyme in rearrangement/isomerase reactions.

COBALT-CORRIN RING COUPLE

The combination of cobalt and the corrin ring has intrigued researchers for a long time and also generates curiosity about the advantages it offers compared to the versatile porphyrin macrocycle found in heme. An equally prompting question is what makes cobalt so well-suited for cobalamin-dependent processes. These questions gain more relevance for two reasons: firstly, the abundance of cobalt is very low and varies significantly in both terrestrial and aquatic environments; secondly, the high degree of complexity associated with the synthesis of the corrin ring.

The driving force behind the incorporation and utilization of cobalt lies in its unique chemical properties. Due to the ability of cobalt to exhibit variable oxidation states, the cobalt ion in cobalamin can exist in different oxidation states including cobalt(I), cobalt(II), and cobalt(III). These oxidation states play a crucial role in imparting reactivity and functionality to cobalamin. Furthermore, the controlled change in the oxidation state of the cobalt ion is intricately related to the chemistry of cobalamin-dependent enzymatic reactions and reflects the oxidation state dependent coordination preferences of the metal ion. For instance, cobalt(III) prefers to bind in a hexa-coordinated manner, whereas cobalt(II) and cobalt(I) adopt penta-coordinated and tetra-coordinated geometries, respectively [9]. These changes in oxidation state are supported by the corrin ring, which in contrast to the planar porphyrin ring, adopts a helical conformation. The binding of the cobalt ion into the corrin ring induces changes in the level of helicity within the corrin structure, depending on the metal oxidation state. This unique characteristic of the corrin ring imparts geometric and electronic adaptability to the molecule for its particular function [9].

Another notable feature is the strong nucleophilicity of the cobalt(I) species, which enables the cobalt ion to form metal-carbon bonds [10]. The ability to control the nature of the Co-carbon complex, particularly in terms of homo- or heterolytic cleavage of this bond, as shown in Figure 5.2, is crucial in the functioning of cobalamin-based enzymes as exemplified in various enzymatic reactions [10].

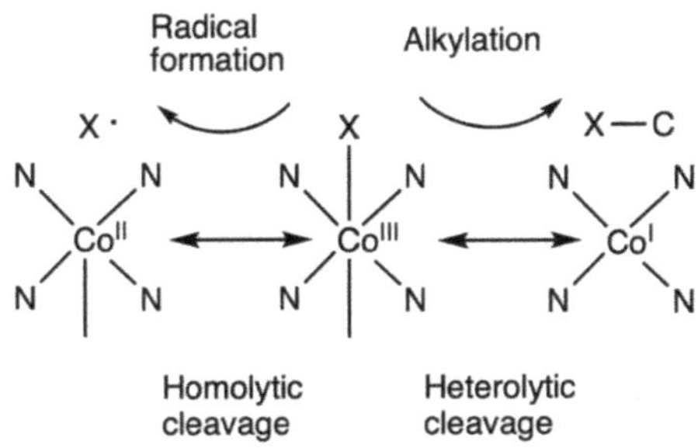

FIGURE 5.2 Schematic representation of hemolytic and heterolytic cleavage in cobalamin derivatives. (This figure is under Creative Common License and is adopted from Ref. [4].)

For instance, methionine synthase employs methyl-cobalamin as a cofactor to catalyze the transfer of a methyl group. The transfer of the methyl group involves the heterolytic cleavage of the Co—CH_3 bond in methylcobalamin, followed by the subsequent transfer of the methyl group to the thiolate of homocysteine, leading to the formation of a nucleophilic Co(I) species. This Co(I) species then readily acquires another methyl group from Me-tetrahydrofolate, generating a Co(III)—CH_3 specie. Notably, the migration of the methyl group occurs with the retention of stereochemistry. The cyclic shuttling of the oxidation is crucial for the efficient transfer of the methyl group from methyl-tetrahydrofolate to homocysteine, leading to the generation of methionine, as shown in Figure 5.3 [7].

On the contrary, isomerization reactions catalyzed by adenosylcobalamin involve homolytic cleavage, which converts the Co(III) into a Co(II) moiety and produces an adenosyl radical. This radical readily abstracts a proton from the substrate, leading to the formation of a substrate radical that subsequently undergoes isomerization. After hydrogen abstraction from the adenosyl group, the Co(III) adenosylcobalamin is regenerated [10]. The generalized reaction mechanism for isomerization reactions, as well as specific examples, is shown in Figure 5.4. Closer inspection of reactions catalyzed by adenosylcobalamin-dependent isomerase revealed that they catalyze a particular type of reaction characterized by a 1,2 interchange of a hydrogen atom and a variable group on adjacent carbon atoms. Broadly, four carbon skeleton isomerases that employ adenosylcobalamin as a cofactor have been identified, as shown in Figure 5.4. These enzymes are crucial for bacterial metabolism. Among them, methylmalonyl-CoA mutase is also found in mammals and facilitates the conversion of catabolites from odd-chain fatty acids, branched-chain amino acids, and cholesterol into succinyl-CoA. Impairment in the activity of methylmalonyl-CoA mutase results in a condition called methylmalonic aciduria, an autosomal recessive disorder characterized by disturbances in secondary acid metabolism [11]. Another intriguing observation is that three out of the four carbon skeleton isomerases mentioned in Figure 5.4 involve the migration of sp^2-hybridized carbon atom, while in glutamate mutase, the migration involves an sp^3 hybridized glycyl moiety. By functioning as a reservoir for radicals, AdoCbl enables the

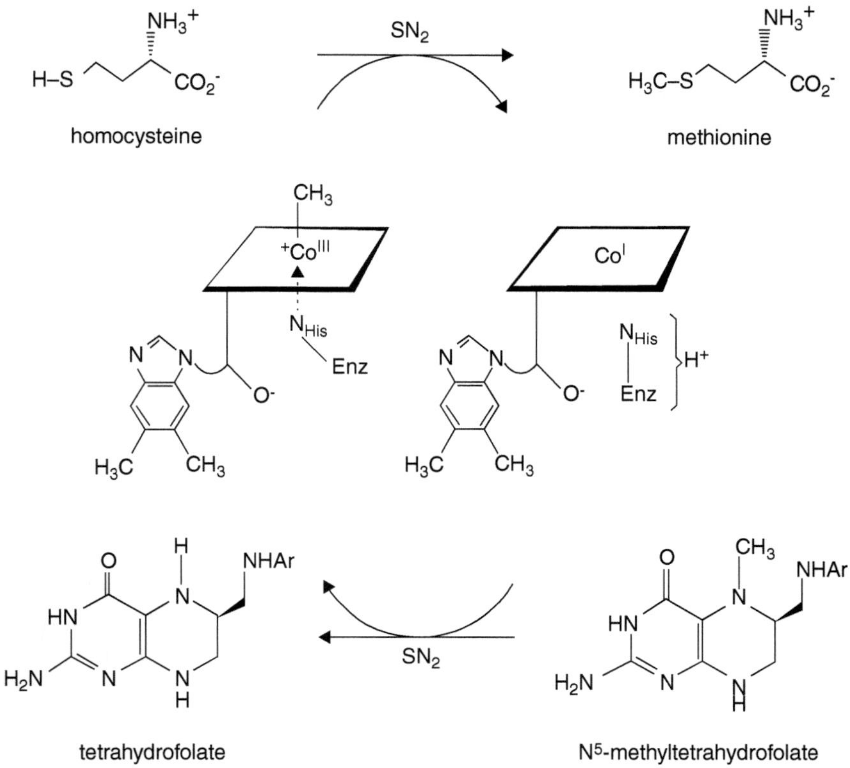

FIGURE 5.3 Methyl transfer reaction mediated by methylcobalamin for the synthesis of methionine. (Copyright permission from Royal Society of Chemistry, Ref. [7].)

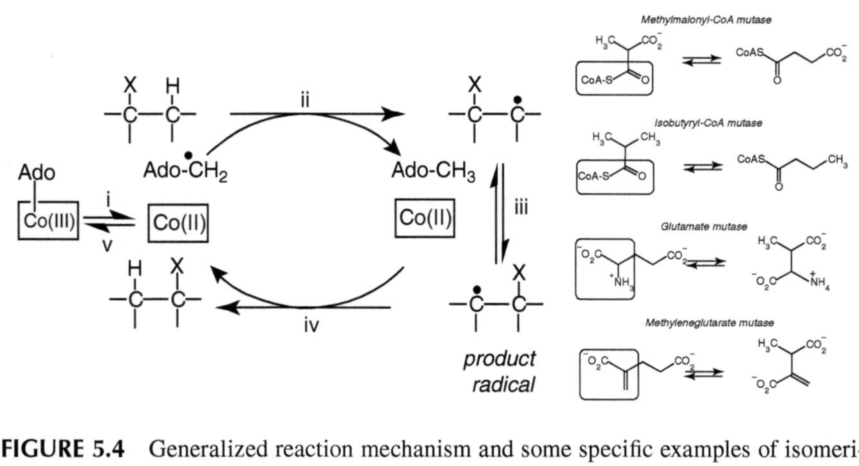

FIGURE 5.4 Generalized reaction mechanism and some specific examples of isomerization reaction catalyzed by adenosylcobalamin-dependent isomerases. (Copyright permission from American Chemical Society, Ref. [11].)

controlled and precise execution of these reactions. Its role as a cofactor in these processes underscores its importance in supporting the intricate chemistry required for these isomerases to catalyze their respective migrations.

Apart from cobalamin-dependent methyltransferases and isomerases, bacteria utilize cobalamin-dependent dehalogenase enzymes as part of their energy metabolism. These dehalogenases are involved in the dehalogenation of aliphatic and aromatic chlorinated hydrocarbons. This process, primarily carried out by anaerobic bacteria, plays a significant role in the detoxification of these compounds [12]. The mechanism of action of dehalogenases involves the formation of an organocobalt adduct, which releases a chloride ion and forms an aryl-Co(III) species. Alternatively, it has been proposed that the corrinoid system donates an electron to hydroxylated aromatics, leading to the formation of a radical anion that subsequently releases a chloride ion. Bacteria couple the dehalogenation reaction to an electron transfer chain, enabling them to derive energy for their metabolic processes [10].

COBALT DEFICIENCY AND TOXICITY

Cobalt, a vital component of vitamin B12, plays a crucial role in various biological processes. Deficiency of cobalt can impede the utilization of vitamin B12 which can have detrimental health effects such as pernicious anemia, neurological disorders, and cognitive impairment. Furthermore, deficiency of cobalt can interfere with DNA synthesis and red blood cell production [13].

On the other hand, increased exposure to cobalt has also been associated with various detrimental health effects. Excess cobalt can trigger contact dermatitis. The American Contact Dermatitis Society recognized cobalt as the "Allergen of the year" in 2016. Acute cases of cobalt toxicity can lead to neurotoxicity, pneumonia, and an increased risk of lung cancer [14].

Most of the adverse effects of cobalt toxicity are linked to the competition between iron and cobalt due to their similarity in size and charge. In bacteria, an excess concentration of cobalt may lead to disruption of iron-sulfur cluster formation and mismetalation of metallo-complexes. It is observed that E. coli, when exposed to 200 μM $CoCl_2$, displays the formation of mixed [Fe/Co-S] clusters in proteins. This interference with iron-sulfur cluster-containing enzymes such as succinate dehydrogenase and fumarate reductase can impede their functionality, leading to metabolic disruptions [4]. Furthermore, cobalt has been found to compete with Fur, the iron regulatory protein, causing disruption of its activity. This may further lead to an imbalance in iron homeostasis, which will have cascading effects on cellular functions [15].

Besides this, cobalt toxicity contributes to the formation of reactive oxygen species. This is attributed to the ability of cobalt to exhibit variable oxidation states. Cobalt in reduced states can react with oxygen and hydrogen peroxide through Fenton-like reactions, resulting in the generation of ROS, superoxides, and hydroxy radicals [16]. These highly reactive species can inflict DNA damage and inhibit DNA repair mechanisms, contributing to cellular dysfunction.

CONCLUSIONS

In summary, this chapter discusses the significant role of cobalt in biology, primarily due to its association with vitamin B12 and its involvement in cobalamin-dependent enzyme activities. The historical perspective of vitamin B12 highlights the Nobel Prize-winning discoveries that led to the identification of cobalt as an integral component of this vital molecule, as well as the structure elucidation and artificial synthesis of vitamin B12. The bioavailability of cobalt is a critical factor in maintaining optimal cobalamin levels in the body, ensuring the proper functioning of cobalamin-dependent enzymes. Furthermore, it is also evident from the chapter that cobalamin-dependent enzymes are involved in various crucial biochemical processes such as methyl transfer, isomerization, and dehalogenation reactions, underscoring their importance in cellular metabolism. These enzymes contribute to essential processes such as DNA synthesis, energy production, and detoxification, highlighting the indispensability of cobalt in biological systems. Imbalances in cobalt levels can lead to severe health issues. Whereas its deficiency is related to impaired function of vitamin B12, leading to pernicious anemia, its excess is also detrimental, causing the disruption of various iron-containing metalloenzymes. Further research in this field will continue to shed light on the intricate mechanisms underlying the role of cobalt in biology and its implications for human health.

REFERENCES

1. L. Florencio, J. A. Field, G. Lettinga, Importance of cobalt for individual trophic groups in an anaerobic methanol-degrading consortium, *Appl. Environ. Microbiol.* 1994, 60, 227–234.
2. Y. Zhang, D. A. Rodionov, M. S. Gelfand, V. N. Gladyshev, Comparative genomic analyses of nickel, cobalt and vitamin B_{12} utilization, *BMC Genomics*, 2009, 10, 78.
3. P. R. Henry, Cobalt bioavailability. In C.B. Ammerman, D.H. Baker, A.J. Lewis (Eds.), *Bioavailability of Nutrients for Animals*, Academic Press, Cambridge, 1995, Pages 119–126.
4. D. Osman, A. Cooke, T. R. Young, E. Deery, N. J. Robinson, M. J. Warren, The requirement for cobalt in vitamin B_{12}: a paradigm for protein metalation, *Biochim. Biophys. Acta-Mol. Cell Res.*, 2021, 1868, 118896.
5. D. C. Hodgkin, J. Kamper, M. Mackay, J. Pickworth, K. N. Trueblood, J. G. White, Structure of vitamin B_{12}, *Nature*, 1956, 178, 64–66.
6. R. B. Woodward, The total synthesis of vitamin B12, *Pure Appl. Chem.*, 1973, 33, 145–178.
7. K. Gruber, B. Puffer, B. Krautler, Vitamin B12-derivatives–enzymecofactors and ligands of proteins and nucleic acids, *Chem. Soc. Rev.*, 2011, 40, 4346–4363.
8. R. Banerjee, C. Gherasim, D. Padovani, The tinker, tailor, soldier in intracellular B12 trafficking, *Curr. Opin. Chem. Biol.* 2009, 13, 484–491.
9. C. Kieninger, E. Deery, A. Lawrence, M. Podewitz, K. Wurst, E. N. Smith, F. J. Winder, A. J. Baker, S. Jockusch, C. R. Kreutz, R. K. Liedl, K. Gruber, M. J. Warren, B. Krautler, The hydrogenobyric acid structure reveals the corrin ligand as an entatic state module empowering B_{12} cofactors for catalysis, *Angew. Chem. Int. Ed. Engl.*, 2019, 58, 10756–10760.

10. R. Banerjee, S. W. Ragsdale, The many faces of vitamin B_{12}: catalysis by cobalamin-dependent enzymes, *Annu. Rev. Biochem.* 2003, 72, 209–247.
11. R. Banerjee, Radical carbon skeleton rearrangements: catalysis by coenzyme B_{12}-dependent mutases, *Chem. Rev.*, 2003, 103, 2083–2094.
12. K. L. Brown, Chemistry and enzymology of vitamin B_{12}, *Chem. Rev.*, 2005, 105, 2075–2149.
13. R. Green, Vitamin B_{12} deficiency from the perspective of a practicing haematologist, *Blood*, 2017, 129, 2603–2611.
14. L. Leyssens, B. Vinck, C. V. D. Straten, F. Wuyts, L. Maes, Cobalt toxicity in humans: a review of the potential sources and systemic health effects, *Toxicology*, 2017, 387, 43–56.
15. C. Ranquet, S. C. –de Ollagnier, L. Loiseau, F. Barras, M. Fontecave, Cobalt stress in Escherichia coli. The effect on the iron-sulfur proteins, *J. Biol Chem.*, 2007, 282, 30442–30451.
16. K. Jomova, M. Valko, Advances in metal-induced oxidative stress and human disease, *Toxicology*, 2011, 283, 65–87.

6 Biochemistry of Nickel

INTRODUCTION

Until relatively recently, nickel had never drawn considerable interest as a biologically important element and only received notable attention due to its toxicity. This adverse property apparently resulted from the binding capability of Ni to DNA through covalent bonds, which in turn caused mutagenic effects. Later on, the need for nickel-mediated processes in anaerobic bacteria for their growth made nickel metal an essential trace bio-element [1,2]. Subsequent findings about nickel metal, as a component of the hydrogenase enzyme, made this belief more promising. Eventually, a close observation of the nature of reactions catalyzed by nickel metal highlighted its undoubted importance for the metabolism of gases like CH_4, CO, and H_2, mainly belonging to pre-oxygen era. This observation also supports the presence of nickel-containing enzymes in a higher number of anaerobic bacteria. The only exception to this series is the urease protein, which is present in bacterial fungi, algae, and higher plants.

Nickel has an electronic configuration of $3d^8\ 4S^2$, which shows oxidation states of 0, 1^+, 2^+, 3^+, and 4^+. Other than the 4^+ oxidation state, all other mentioned oxidation states are reported in biological systems. The most common oxidation state of nickel is +2, with a square planar geometry. It forms a stable complex after copper (II) among first-row transition elements, with amino acid residues like histidine (log $K_f = 15.9$) [1]. To attain other oxidation states, e.g., Ni(0), strong field ligands like carbonyl, cyanide, etc., are required. However, Ni(III) and Ni(IV) can be stabilized by the presence of ligands with hard donor atoms, such as oxygen and nitrogen (Table 6.1). The low oxidation state of the metal exposes some of its

TABLE 6.1
Biological Nickel and Their Geometry in Various Coordination Number

S.N.	Ni	C.N.	Nature	Magnetic Nature	Geometry
1	Ni(O)	4	Low spin	Diamagnetic	Tetrahedral
		4	Low spin	Diamagnetic	Square planar
2	Ni(I)	4	High spin	Paramagnetic	Square planar
		6	High spin	Paramagnetic	Distorted octahedral
3	Ni(II)	4	Low spin	Diamagnetic	Square planar
		4	High spin	Paramagnetic	Tetrahedral
		5	High spin	Diamagnetic	Square pyramidal
		6	High spin	Paramagnetic	Octahedral
4	Ni(III)	5	Low spin	Paramagnetic	Trigonal bipyramidal
		6	Low spin	Paramagnetic	Octahedral

DOI: 10.1201/9781003432333-6

TABLE 6.2
Nickel Enzymes and Their Role in Various Biological Reactions

S.N.	Enzyme	Catalytic Reaction	Organism
1	Urease	Urea + $H_2O \rightarrow 2NH_3 + H_2CO_3$	Archaea, bacteria
2	Hydrogenase	$2H^+ + 2e^- \rightarrow H_2$	Archaea, bacteria
3	CO dehydrogenase	$CO + H_2O \rightarrow 2H^+ + CO_2 + 2e-$	Archaea, bacteria
4	Methyl-Co-M reductase	Methyl-SCoM + CoBSH $\rightarrow$ CH_4 + CoBS-SCoM	Bacteria
5	Acetyl-Co-A synthase	CH_3CFeSP + CoASH + CO $\rightarrow$ CH_3–CO–SCoA + CFeSP	Archaea, bacteria
6	Glyoxalase	Methylglyoxal + glutathione $\rightarrow$ GS-derivative	Archaea, bacteria
7	Acireductone dioxygenase	Aci-reductone + $O_2 \rightarrow$ methylthiopropionate	Archaea
8	Ni superoxide dismutase	$2O_2^- + 2H^+ \rightarrow H_2O_2 + O_2$	Bacteria

d electrons to form σ or π orbitals, which afford tetragonal reactive nickel compounds, similar to a free radical moiety. Ni(O) is especially facilitated easily in the catalytic cycles of hydrogenase and acetyl-CoA synthase, etc. Furthermore, nickel metal displays diamagnetic behavior only in square planar geometry and paramagnetic behavior in tetrahedral, square pyramidal, and octahedral structures. Conclusively, nickel complexes have a wide range of geometries and corresponding oxidation states depending on the nature of the coordinating ligand system. It is expected that the protein environment around biological nickel would effectively govern its properties.

Only eight nickel-containing enzymes, namely urease (involved in ammonia formation), acireductone dioxygenase (ARD) (involved in CO formation), hydrogenase (produces or utilizes H_2), CO dehydrogenase (interconverts CO and CO_2), acetyl-CoA synthetase (converts CO_2), methyl-CoM reductase (MCR) (produces CH_4), SOD (generates O_2 from superoxide), glyoxalase (GlxI) (catalyzes the conversion of toxic methylglyoxal to lactate), are reported to date (Table 6.2). Among them, seven are involved in the consumption or production of reducing gases, such as CH_4, NH_3, CO, CO_2, and H_2. In other words, we can say these enzymes play a pivotal role in the biological cycles of carbon, nitrogen, and oxygen. These nickel-based enzymes generally consist of several protein units (multimeric) and are classified into four classes: ureases, CO dehydrogenases/acetyl-CoA synthetases (CODH/ACS), S-methyl coenzyme M reductases, and hydrogenases (Table 6.3). There may still be the possibility of discovering many additional Ni-enzymes, which will enhance the role of biological nickel [1].

UREASE

Urease is the very first nickel-containing enzyme isolated from jack beans. In contrast to R. Willstatter's view, the urease enzyme was crystallized by James Sumner in 1926 but the existence of nickel in the enzyme came to light 50 years later [2,3].

TABLE 6.3
Characteristic of Representative Ni Based Enzymes

S.N.	Protein	Mol. Wt.	Cluster Composition
1	Urease	α=60.3, β=11.6, 11.0 KDa	2 Ni ions
2	Hydrogenase	L=58 KDa, S=32 KDa	NiFe and 2[Fe_4S_4]
3	CODH	α=867 KDa β=72.3 KDa	A cluster [NiFe], B cluster [Ni_4Fe_4] C cluster [NiFe]
4	ACS	78 kDa	$NiNiFe_4S_4$ cluster
5	Methyl-CoM reductase	65 (Ni protein), 46 and 35 KDa	Ni tetrapyrrole
6	Superoxide dismutase	-----	4–5 coordinate Ni site
7	Glyoxalase I	14.9 KDa	6-coordinate Ni site
8	Aci-reductone dioxygenase	20.2 KDa	6-coordinate Ni site

Urease enzymes catalyzes the hydrolysis of urea to carbamate which spontaneously hydrolyzes to carbonic acid as well as the second molecule of ammonia (Equations 6.1 and 6.2).

$$\mathbf{NH_2CONH_2 + H_2O \longrightarrow NH_2COOH + NH_3} \quad (6.1)$$

$$\mathbf{NH_2COOH + H_2O \longrightarrow H_2CO_3 + NH_3} \quad (6.2)$$

In 1915, Richard Willstatter, who received the Nobel Prize in Chemistry, proposed that proteins were not enzymes and that urease enzymes were simply scaffolds for catalysis. Later, he was proven wrong by James Sumner, who earned the Nobel Prize in 1946 for 'Crystallization of protein'[3] (Figure 6.1).

In urease, two non-equivalent nickel centers are present. The geometry of Ni1 is pseudo-tetrahedral with a weakly occupied fourth ligand. However, Ni2 defined a square pyramidal or trigonal bipyramidal geometry with a coordination number of five. Both nickel centers are bridged by carbamylated lysine residues. Tetrahedral oriented N1 has coordinated with two histidine, lysine and weakly bound water whereas Ni2 had coordination of two histidine, one aspartate and water. Each nickel exists in a +2 oxidation state with diamagnetic behavior. In general, Ni(II) compounds coordinated with O and N ligands tend to achieve octahedral geometry. This clearly indicates the influence of a defined protein environment on the nickel coordination geometry in the urease enzyme [2,3] (Figure 6.2).

Mechanism

Urea is a stable compound that hydrolyzes to isocyanic acid and ammonia. The half-life of this uncatalyzed reaction is 3.6 years.

$$NH_2CONH_2 + H_2O \rightarrow 2NH_3 + CO_2 \quad (6.3)$$

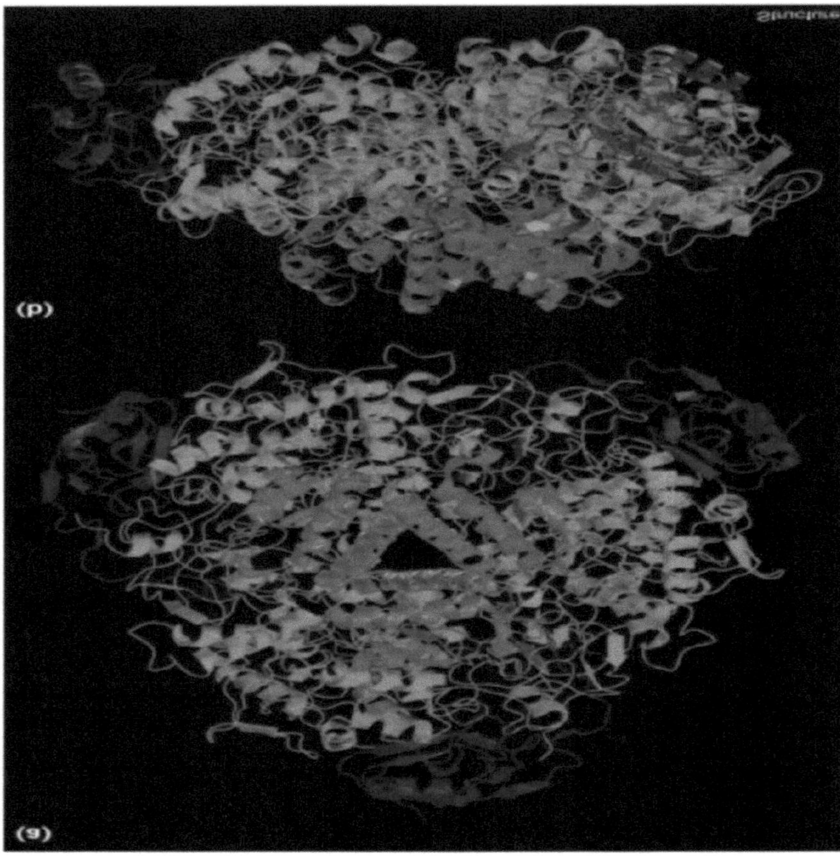

FIGURE 6.1 Ribbon diagram of the enzyme structure of urease from *Bacillus pasteurii*. The blue, light blue, and light grey ribbons represent the α, β, and γ subunits, respectively. The Ni ions are located at the top of the α subunit. (Reused with permission from Ref. [2]; Copyright (2023) Elsevier.)

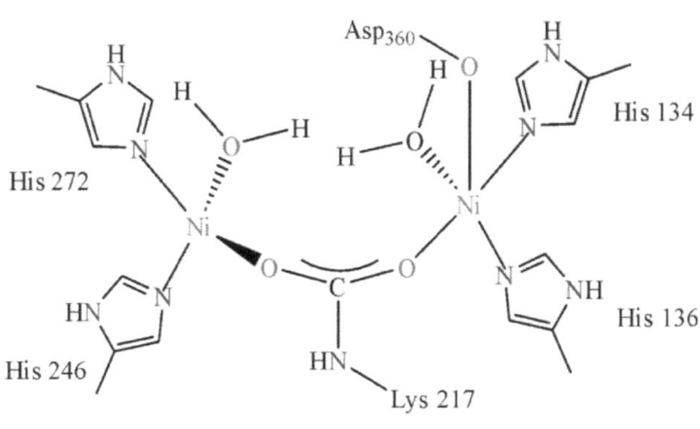

FIGURE 6.2 The active site of urease enzyme with two nickel ions.

However, as mentioned earlier, some reaction in the presence of urease enzymes afforded two moles of ammonia along with carbon dioxide (Equation 6.3). The rate of completion in the presence of urease enzyme is increased by 10^{14} times. The acceleration could be explained by looking into the insights of the mechanism. In the first step, polarization of the substrate (urea) takes place through the binding of the oxygen atom to the Ni1 center, followed by the subsequent nucleophilic attack of the Ni-bound hydroxyl group (formed by proton abstraction by His 320 from the water molecule attached to the Ni2 center). The terminal $-NH_2$ available on cyclic transition state, upon protonation from His 320, would eliminate the NH_3 molecule, leaving the carbamate bridged between two Ni atoms [2,3] (Scheme 6.1). Dissociation of carbamate from the binuclear site would result in spontaneous hydrolysis to carbonate and a second molecule of NH_3, with a protonated His residue acting as an acid to promote ammonia release. Conclusively, the main points of the mechanism responsible for lowering down the kinetic barrier of urea hydrolysis reactions are:

1. Coordination of urea to the Ni1 metal center results in the carbonyl group being more electrophilic.
2. Formation of an activated nucleophilic nickel hydroxyl species.
3. Stabilization of all $-NH_2$ groups to urea through hydrogen bonding.

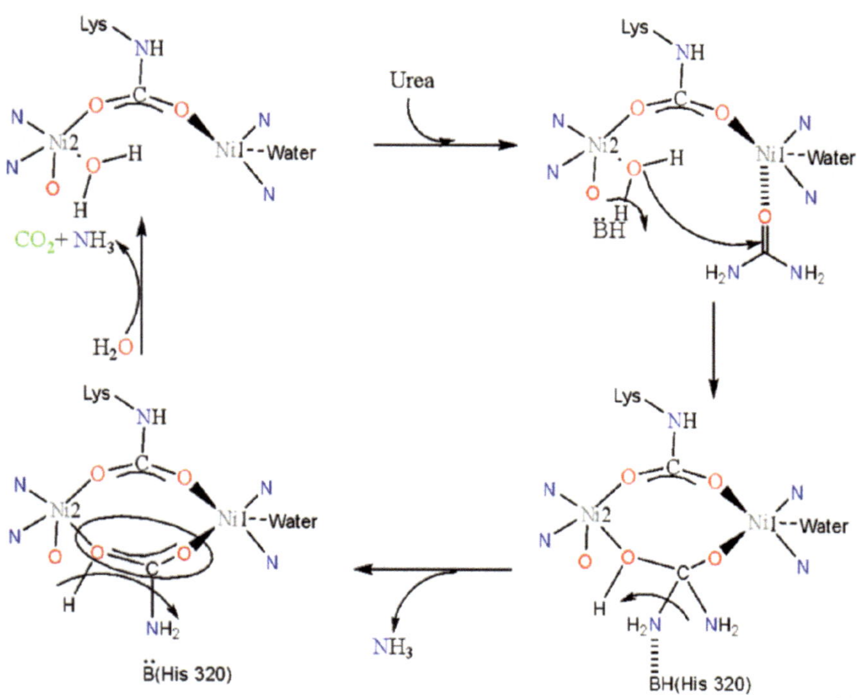

SCHEME 6.1 Plausible catalytic cycle for urease enzyme.

HYDROGENASE

Hydrogenase enzyme catalyzes the generation or splitting of hydrogen molecules as per the following reaction:

$$H_2 \rightarrow 2H + 2e^- \tag{6.4}$$

The reaction is heterolytic in nature with the formation of hydride (H^-) as well as (H^+) in the preliminary step. The production of hydrogen is required to avoid the buildup of the reduction potential in anaerobic bacteria; however, its splitting is concerted with the reduction of CO_2, CO, other electron acceptors, etc., and is used to generate a proton gradient for ATP production requirements [4]. Three types of hydrogenases are reported: (i) [Ni-Fe] hydrogenase, (ii) [Fe_3-Fe] hydrogenase, and (iii) [Fe] hydrogenase. Since our main limitation here is to focus on nickel-mediated hydrogenase, we will discuss only the [Ni-Fe] hydrogenase.

The crystal structure of *D. gigas* hydrogenase showed a hetero-binuclear core of Ni and Fe metal centers, present in two subunits, i.e., large and small. The large subunit contains three iron-sulfur clusters and an active nickel center present in the larger subunit. The nickel center is surrounded by four cysteine residues, two of which are terminal, and the remaining two act as a bridge between Ni and Fe centers (Figure 6.3). Protein ligation corresponds to the square pyramidal geometry around the nickel center with an in-plane vacant ligand site. However, the crystal structure of oxidized enzymes shows that the vacant site is occupied by putative incoming ligands, which assist in completing the expected square pyramidal coordination around the nickel metal. Spectroscopic and crystallographic studies indicate the presence of three non-protein diatomic molecules, i.e., CO and cyanide ligands. The presence of these strong ligands forces the iron into a low-spin diamagnetic state, i.e., the ferrous state [4]. With the presence of six ligands, iron in the oxidized form adopts an octahedral arrangement. The implications of the unique [Ni-Fe] core on H_2 catalysis are discussed below.

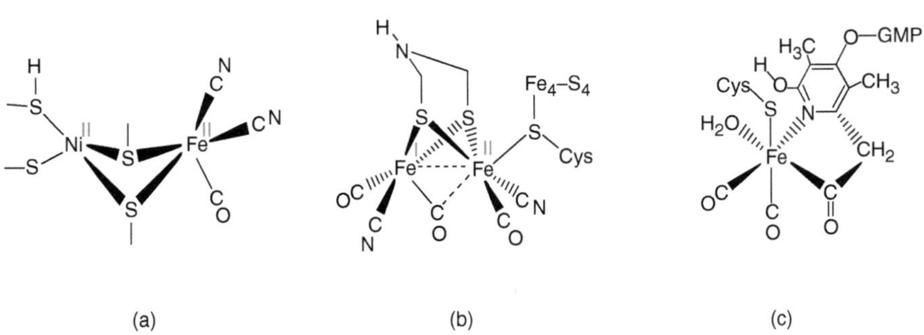

FIGURE 6.3 Active site structures of three phylogenetic hydrogenases: (a) [NiFe]-hydrogenase from *D. gigas*, (b) [FeFe]-hydrogenase from *C. pasteurianum*, and (c) [Fe]-hydrogenase from *M. jannaschii*. (Reused with permission from Ref [3]; Copyright (2023) Elsevier.)

TABLE 6.4
Biological Nickel with Their 'g' Values

Enzyme	Core	Type of Ni with g Values	Comment
Hydrogenase	[NiFe]	Ni_A=2.32, 2.23, 2.01	Unready form
		Ni_B=2.34, 2.16, 2.01	Ready form
		Ni_C=2.35, 2.16, 2.01	Active form

Mechanism

In general, [NiFe] hydrogenase needs to be activated by prolonged treatment with hydrogen. During this proposition, the nickel ion serves as the catalytic site where heterolytic splitting of H_2 takes place into H^+ and H^-, and nickel undergoes several changes in oxidation from Ni(I) to Ni (III), and categorized into three phase: Ni_A, Ni_B, and Ni_c, depending on their functioning with respect to g values (Table 6.4). In the very first phase, the thiolate of Cys530 plays a role in base catalysis upon conversion of Ni_B(III) to Ni_C(I) through hydride transfer or proton-coupled electron transfer reactions (Scheme 6.2). Electron-rich Ni(I) becomes susceptible to binding of H_2. H–H bond cleavage is proposed to occur by an oxidative addition reaction mechanism that generates the Nia-X*(B, III) intermediate [1,4]. Subsequently, two successive proton-coupled electron transfer steps rejuvenate the diamagnetic Ni(II) state. A plausible catalytic cycle based on electron transfer can be postulated as follows.

CODHs

CODH is the enzyme that facilitates CO as a carbon and electron source for various microorganisms. It catalyzes the oxidation of CO with an effective conversion rate of 4,000–40,000 s^{-1} depending on the source, and also reduces the CO_2 at a rate of 11 s^{-1} under mild conditions. Two types of Ni-CODH are reported: the first is a homodimeric uni-functional CODH present in anaerobic bacteria such as *Rhodospirillum rubrum* and *Carboxydothermus hydrogenoformans*. These types of enzymes are coupled with Co-induced hydrogenase, which oxidizes CO to CO_2 while forming hydrogen [5,7]. This assists the growth of organisms using CO as an energy source. The second type, bifunctional, contains CODH as well as acetyl Co-A synthase, found in acetogenic bacteria such as *Moorella thermoacetica*. These enzymes catalyze, the synthesis of acetyl-CoA, allowing the involvement of CO and CO_2 through the Wood-Ljungdahl fixation pathway (Figure 6.4).

Carbon-Monoxide Dehydrogenase (CODH)

Structure

The crystal structure of unifunctional CODH reveals a mushroom-shaped homomeric protein with a molecular weight of 130 KDa. The dimer contains five metal-sulfur clusters, i.e., two B, two C, and D. Electrons generated during the

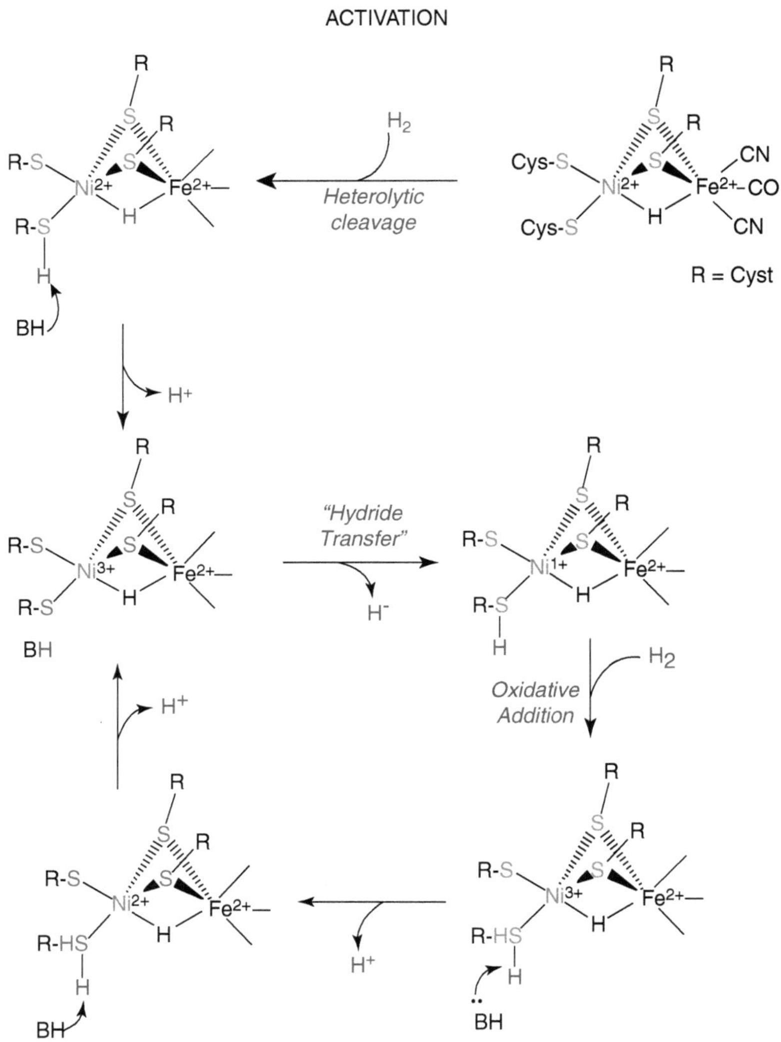

SCHEME 6.2 Plausible catalytic cycle for hydrogenase enzyme. The asterisks indicate an EPR-active state. (Redrawn from Ref. [5].)

process are transferred to a wire consisting of cluster B, a typical [4Fe-4S] cluster, and cluster D, which has a [4Fe-4S] composition with two cysteine residues. An active center is usually present in the 'C' cluster, which consists of a unique asymmetric assembly [$NiFe_4S_5$], where nickel is tetra-coordinated with a 4S-defined core having square planar geometry (Figure 6.5). One of the iron atoms is extraneous to the cuboidal core and is connected to nickel through a bridging sulfide [6,7]. However, in non-reduced CODH, nickel is present at the cuboidal corner with a pentacoordinate square pyramidal geometry, where the fifth ligand is allocated as 'CO'. The iron atom is connected through bridging cysteinyl sulfur rather than sulfide.

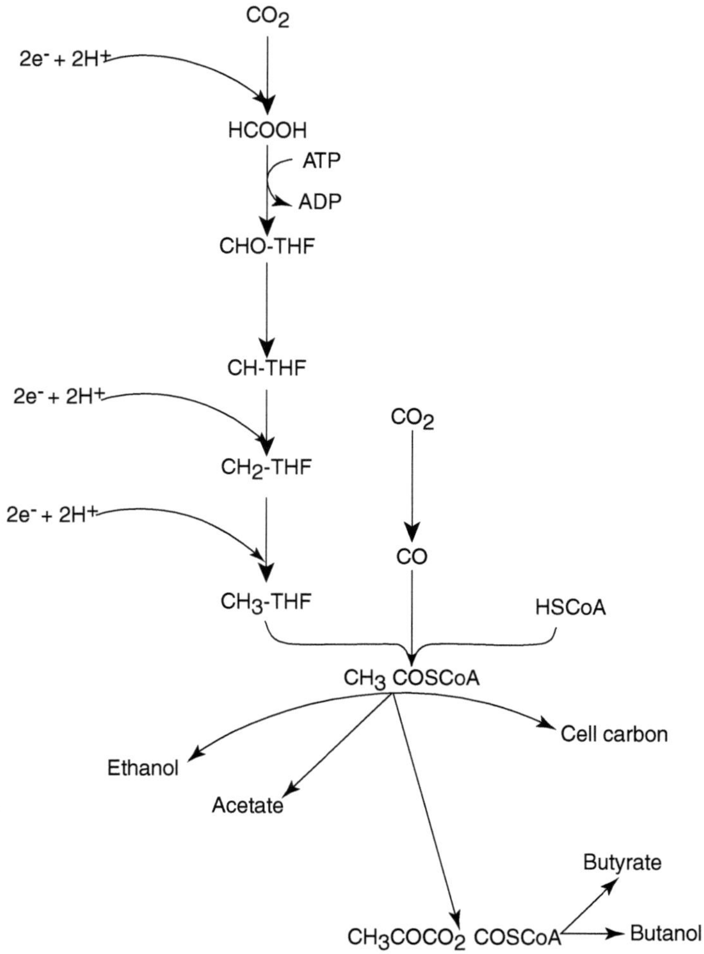

FIGURE 6.4 Wood-Ljungdahl fixation pathway used in autotrophic growth in anaerobic bacteria. (Adapted from Ref. [9].)

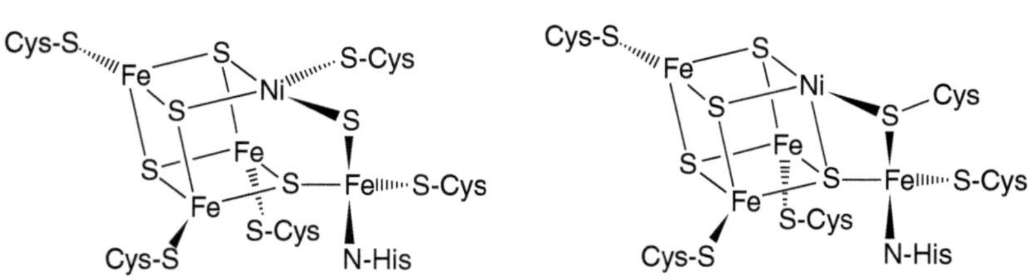

FIGURE 6.5 The structures of CODH C-cluster: (a) reduced form in *C. hydrogenoformans*; (b) non-reduced form in *R. rubrum*. (Reused with permission from Ref. [8]; Copyright (2023) Elsevier.)

Mechanism

To explore the mechanistic prospects involved in CODH function, a catalytic cycle has been proposed as follows (Scheme 6.3):

1. Water binds to the extraneous iron atoms of the cluster, where it gets deprotonated to yield a nucleophilic hydroxide ion on the iron center.
2. Nickel metal is coordinated to CO, followed by the nucleophilic attack of the hydroxide ion to carbonyl carbon to afford carboxylic acid group on nickel atom.
3. The resultant carboxylic group bound to nickel is deprotonated and escapes through channels, returning nickel to its zero oxidation state.

Delivering two electrons to the active center rejuvenates the enzyme active site for another cycle.

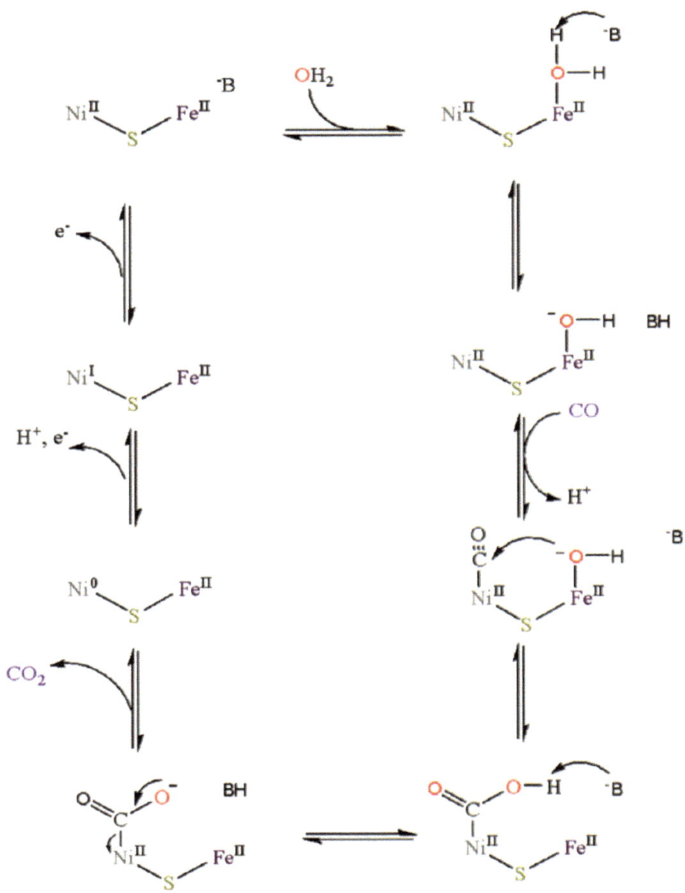

SCHEME 6.3 Plausible catalytic cycle of CODH enzyme. (Redrawn from Ref. [7].)

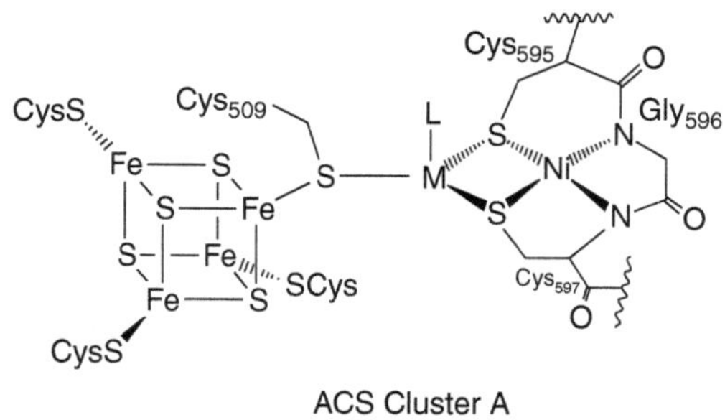

FIGURE 6.6 Crystal structure of cluster A of ACS enzyme. (Reused with permission from Ref. [8]; Copyright (2023) Elsevier.)

ACETYL-COA SYNTHASE (ACS)

STRUCTURE

This bifunctional enzyme consists of a tetrameric composition, $\alpha_2\beta_2$, with a molar mass of 310 KDa. The β domain structure is similar to that of unifunctional CODH. The α domain contains cluster A, which is responsible for ACS activity [7–10]. To execute the basic function of CODH/ACS, a hydrophobic channel is present between the C cluster in CODH and the A cluster in the ACS subunit (Figure 6.6). This tunnel couples CO production and facilitates its release from one active site to the other, connecting A to cluster C. Cluster A is composed of $[Fe_4S_4]$ cluster, bridged through cysteinyl sulfur to a proximal metal, and is then connected to a distant Ni(II) atom with a defined coordination core 'N_2S_2'. Earlier studies show the presence of tetrahedral Cu(I) at the proximal end. Later on, tetrahedral Zn(II) was assigned in the closed protein conformation, and square planar Ni(II) is present in the open conformation of the protein [9,10]. The functional monomeric ACS contains $[Fe_4S_4]$-Ni-Ni assembly. Basically, an enzymatically active A cluster contains a proximal nickel atom, in which the fourth ligand is still unidentified but is labile in nature. Under adverse conditions, substitution with Cu and Zn renders it inactive.

MECHANISM

ACS catalyzes the final step of a biological process called the Wood-Ljungdahl process, in which a methyl group, CO, and CoA are clubbed to form Acetyl-CoA. Two different mechanisms (Schemes 6.4 and 6.5) have been suggested to understand the functioning of the ACS enzyme. Both mechanisms initiate in a similar way, with the coordination of CO in the first step [9,10]. The possible variations rise from the oxidation state of the proximal nickel center. The detailed scenario of the mechanism is as follows.

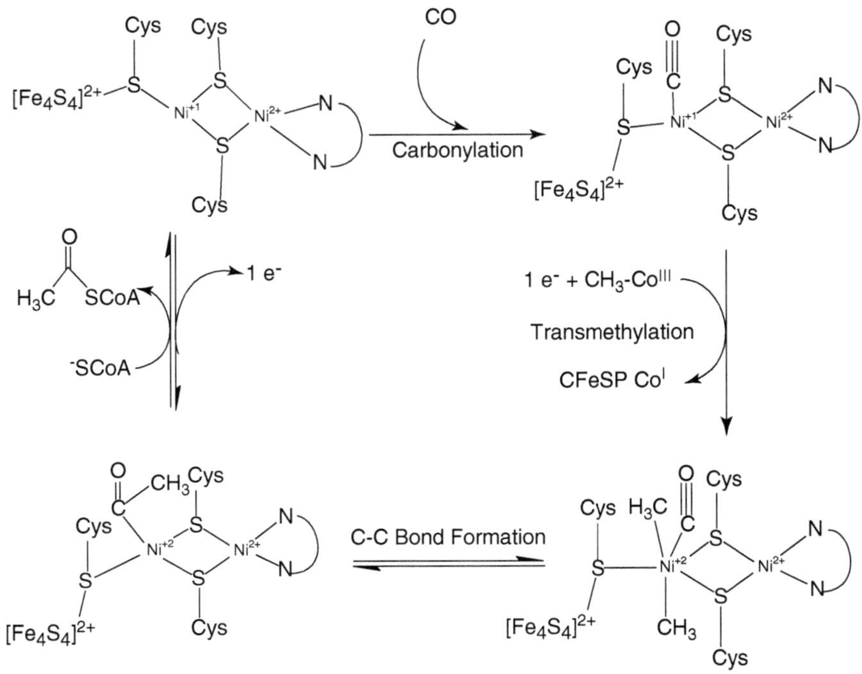

SCHEME 6.4 Plausible catalytic cycle 1 for ACS enzyme scheme.

Mechanism 1

1. In this proposed cycle, the 'CO' moiety binds to the paramagnetic Ni(I) site. The paramagnetic nature of the nickel center and the axial orientation of CO on the intermediate NiFeC species are verified by IR and EPR studies (Scheme 6.4).
2. Nucleophilic attack of cluster A on the methyl group of methylated CFeSP forms the methylated state of the A cluster [8]. To make the nickel center diamagnetic, an internal electron transfer reaction involving Fe-S takes place.
3. Either insertion of CO into the Ni^{2+}–CH_3 bond or migratory substitution of the methyl group to Ni-CO bond results in carbon-carbon bond formation, yielding the acetyl metal species.

At the last step, deprotonated CoA binds and removes the acetyl moiety as Acetyl-CoA. The single electron transformed during the methylation is returned through the channel to regenerate the Ni(I) active site (Scheme 6.4).

Mechanism 2

1. At the first stage, binding of CO takes place to the proximal Ni(O) atom, which is located beneath the gas channel [7,8] (Scheme 6.5).

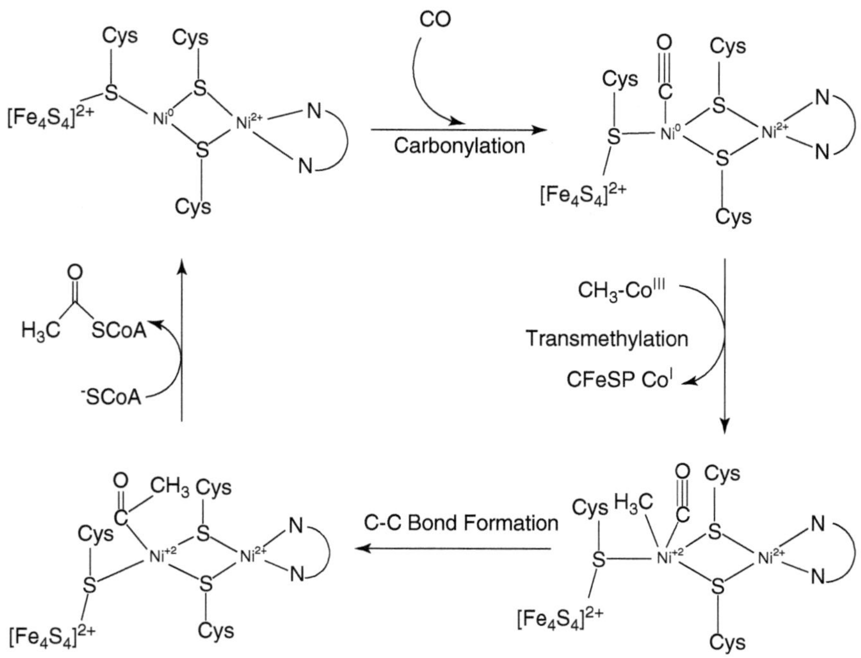

SCHEME 6.5 Plausible catalytic cycle 2 for ACS enzyme.

2. The second step involves the nucleophilic attack of cluster A on methylated-corrinoid-iron-sulfur, resulting in methylated cluster A. The transfer of the methyl group takes place at the same nickel, while it gains two electrons.
3. The axially oriented 'CO' group inserts into the Ni-methyl bond and facilitates carbon-carbon bond formation, which leads to the formation of the nickel acetyl species.
4. In the last step, the deprotonated CoA group attacks the nickel acetyl moiety to form the acetyl-CoA product. This process liberates two electrons, which help redevelop the Ni(O) state. The $[Fe_4S_4]$ cluster plays no significant role in the redox activities of the catalytic cycle [9,10] (Scheme 6.5).

It is quite common for ACS to assume two conformations: one in which the A cluster is not accessible, and the CO channel is accessible to the solvent, and a second one in which the A cluster is accessible, and the CO channel is inaccessible to the solvent [6–8]. It is also important to note that major conformational changes occur during acetyl-CoA synthesis.

CONCLUSIONS

Four important roles of biological nickel, i.e., the hydrolysis of urea in urease, the splitting of hydrogen in hydrogenase, the oxidation of CO, and the synthesis of acetyl-CoA in CODH, have been discussed. Among them, urease is considered the simplest, as no redox chemistry has been observed, and only the stable Ni(II) state is present during the course of the reaction. However, moving from urease to hydrogenase or CODH, their functioning becomes complicated due to the involvement of extensive redox behavior of the nickel center. Hence, the mechanisms of these enzymes have been subject to differing opinions. The existence of nickel in hydrogenase prior to iron could have been possible because of the redox potential range of the nickel complex. Established information related to various oxidation potentials, states, binding energies, etc., helps to understand the mechanisms and functioning of these enzymes.

REFERENCES

1. R. R. Crichton, *Biological Inorganic Chemistry: A New Introduction to Molecular Structure and Function* (Second Edition), Elsevier, Amsterdam, The Netherlands, 2012, Pages 297–310.
2. S. Benini, W. R. Rypniewski, K. S. Wilson, S. Miletti, S. Ciurli, S. Mangani, A new proposal for urease mechanism based on the crystal structures of the native and inhibited enzyme from Bacillus pasteurii: why urea hydrolysis costs two nickels, *Structure*, 1999, 7, 205–216.
3. D. M. Heinekey, Hydrogenase enzymes: recent structural studies and active site models, *J. Organomet. Chem.* 2009, 694, 2671–2680.
4. S. Shima, O. Pilak, S. Vogt, M. Schick, M. S. Stagni, W. M.- Klaucke, E. Warkentin, R. K. Thauer, U. Ermler, The crystal structure of [Fe]-hydrogenase reveals the geometry of the active site, *Science* 2008, 321, 572–575.
5. S. W. Ragsdale, Nickel-based enzyme systems, *J. Bio. Chem.* 2009, 284, 18571–18575.
6. S. W. Ragsdale, Nickel enzymes & cofactors. In Robert A. Scott (Ed.), *Encyclopedia of Inorganic and Bioinorganic Chemistry*, University of Michigan, New Jersey, 2011, Pages 1–16.
7. D. J. Evans, Chemistry relating to the nickel enzymes CODH and ACS, *Coord. Chem. Rev.* 2005, 249, 1582–1595.
8. C. G. Riordan, Bioorganometallic chemistry of cobalt and nickel. In J. McCleverty, T. J. Meyer (Eds.), Bioorganometallic Chemistry of Cobalt and Nickel, *Comprehensive Coordination Chemistry* (Volume Eighth), Elsevier, Oxford, 2003, Pages 677–713.
9. O. T.- Acevedo, M. S. Chinn, A. M. Grunden, Production of biofuels from synthesis gas using microbial catalysts, *Adv. Appl. Microbiol.* 2010, 70, 57–92.
10. W. Kaim, B. Schwederski, A. Klein, *Bioinorganic Chemistry: Inorganic Elements in the Chemistry of Life* (Second Edition), John Wiley & Sons, New Jersey, USA, 2013, Pages 1–425.

7 Bioinorganic Prospect of Copper-Containing Enzymes

INTRODUCTION

Many copper-containing metalloenzymes are present in both prokaryotes and eukaryotes. These copper analogues have parallel characteristics to iron-containing proteins in terms of comparative functions, such as electron transfer reactions (azurin, plastocyanin), oxygenation reactions (tyrosinase and ascorbate oxidase), and oxygen transport (hemocyanin) [1] (Table 7.1). Despite these similarities, they differ in their coordination environment, physiological appearances, and functional prospects. The utility of iron-containing protein enzymes in higher organisms is greater; however, copper-based biomolecules are well reported in both primitive and post-primitive eras. The basic differences between iron- and copper-based proteins are listed in Table 7.2.

Most of the copper-containing proteins occur outside the cell. However, the enzyme superoxide dismutase exists in both cytosols and inside the cell. Copper present in these biomolecules show three different oxidation states, i.e., 1^+, 2^+, and 3^+ oxidation states with electronic configuration d^{10}, d^9, and d^8 electronic configuration, respectively (Table 7.3) [2]. Copper (I) d^{10} exists in 2, 3, or 4 coordination numbers with linear, trigonal planar, pyramidal, or tetrahedral geometry, respectively (Table 7.4). However, in the case of Cu(II), the asymmetric distribution of electrons in higher energy levels exhibit John-Teller distortion to offer favorable tetragonal geometry. The d^8 configured Cu(III) favors square planar geometry. The Irving-William series indicates that, among the first row of transition metal

TABLE 7.1
Corelation between Iron and Copper Proteins

S.N.	Function	Iron Protein	Copper Protein
1	Oxygen transport	Hemoglobin and hemerythrin	Hemocyanin
2	Oxygenation	Cytochrome P-450 and catechol dioxygenase	Tyrosinase and dioxygenase
3	Oxidase activity	Peroxidase (heme and non-heme)	Amine oxidase and laccase
4	Electron transfer	Cytochrome	Azurin and plastocyanin
5	Antioxidant	Bacterial superoxide dismutase	Cu-Zn superoxide dismutase
6	NO_2^- reduction	Heme-containing nitrite reductase	Cu-containing nitrite reductase

DOI: 10.1201/9781003432333-7

TABLE 7.2
Differences between Biological Copper and Iron

S.N.	Biological Copper	Biological Iron
1	Biological copper is mostly bonded to nitrogen through a strong and kinetically inert bond	Iron is coordinated to a tetrapyrrole macrocyclic ligand
2	Redox pairs of $Cu^{I/II}$ have a higher redox potential compared to $Fe^{II/III}$. Hence, ceruloplasmin protein catalyzes the oxidation of $Fe^{II/III}$ (ferroxidase)	Any such activity has been noted in case of iron in biological system
3[a]	The Cu^{2+} form is more soluble compared to the reduced Cu^{1+}. As a result, the oxidized form in results insoluble compounds with halides and sulfur	Poor solubility of Fe^{3+} ions has an impact on more iron precipitation during early evolution in terms of a geochemical approach
4	Copper is present in the extracellular volume or vesicle	Iron is mostly present inside the cell

[a] In geochemical implications, considering the higher solubility of Cu^{2+} compared to Fe^{3+} derives iron precipitation and copper mobilization during the primitive era of biogenic oxygen products.

TABLE 7.3
Nature of Copper Metal with Respect to Oxidation State

S.N.	Oxidation State	Electronic Configuration	Magnetic Properties	Nature Hard/Soft	Favorable Amino Acid
1	Cu^{1+}	$3d^{10}$	Diamagnetic	Soft	Cyst., Met.
2	Cu^{2+}	$3d^9$	Paramagnetic	Hard	H_2O, His, Tyr, Ser, Thr.
3	Cu^{3+}[a]	$3d^8$	Diamagnetic	Hardest	Tyr^-, OH^-

[a] Cu^{3+} would be reduced to Cu^{2+} by soft ligands.

TABLE 7.4
Favored Geometry of Copper in Various Coordination Number

S.N.	Oxidation State	Coordination Number	Geometry
1	Cu^{1+}	2	Linear
		3	Trigonal planar
		4	Tetrahedral
2	Cu^{2+}	4	Square planar
		4	Tetragonal
		5	Square pyramidal
		6	Distorted octahedral
3	Cu^{3+}	4	Square planar

series, Cu^{2+} forms the most stable complexes. This Cu^{2+} is hard in nature and prefers the nitrogen ligands, whereas soft copper (I) prefers sulfur-containing ligands as well as ligands with more polarizable electron clouds, e.g., o-phenanthroline (Table 7.3) [2,3]. In general, copper-containing biomolecules have been classified into three categories depending upon structural and spectroscopic analysis as proposed by Malkin and Malmstrom. These are as follows:

1. Type 1 2. Type 2 3. Type 3

Type 1: Copper enzymes are also called blue copper proteins, having tetrahedral geometry (in plastocyanin) or trigonal bipyramidal geometry (in azurin), depending on the nature of the protein. Commonly, type 1 copper is ligated with histidine, one cysteine, and one methionine residue. Recent research confirmed the presence of carbonyl oxygen from amide as a fifth ligand in some species of azurin. These proteins are blue in nature due to a characteristic absorption band at 600 nm in the visible region, with an extinction coefficient ≥ 3,000 mol/cm. One can easily imagine the intensity of the color by comparing it with blue $[Cu(H_2O)_6]^{2+}$ (Figure 7.1), which has an extinction coefficient of 5–10 mol/cm.

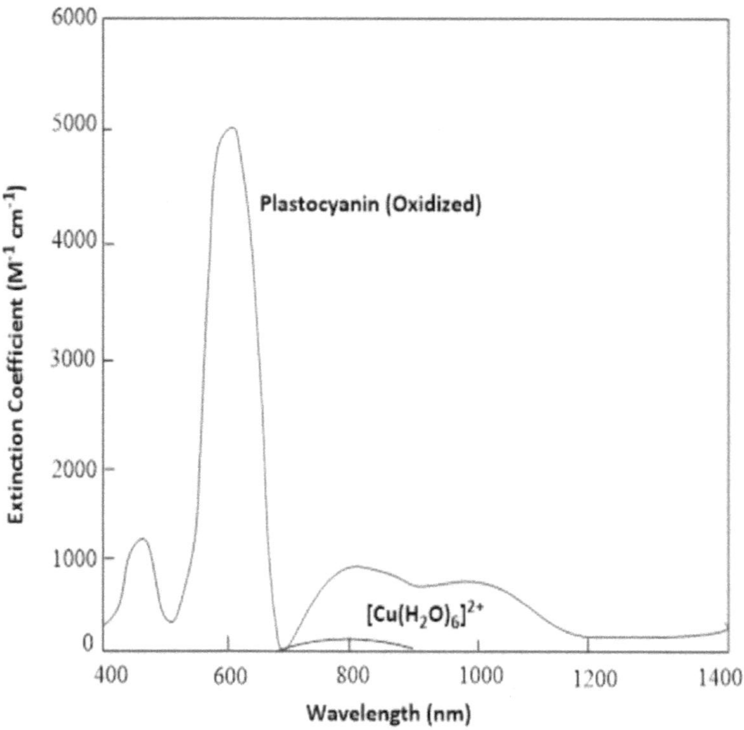

FIGURE 7.1 Comparison of electronic spectra between oxidized plastocyanin and $[Cu(H_2O)_6]^{2+}$.

TABLE 7.5
Classification of Copper Enzymes

Type	Enzyme	Mol.Wt. (kDa)	Absorption Band (nm)	EPR (Hyperfine Coupling; $A \times 10^{-4} cm^{-1}$)
Type 1	Azurin	14	625	60
	Plastocyanin	10.5	597	50
	Laccase	60–140	607	63
	Ascorbate oxidase	145	614	43
Type 2	Laccase	60–90	760	199
	Ascorbate Oxidase	145	788	200
Type 3	Tyrosinase	46	345 and 590	EPR silent
	Hemocyanin	400–900	350 and 570	EPR silent

The absorbance band at 600 nm arises due to the electronic transition from Cys to Cu^{2+}, a ligand-to-metal charge transfer band [1]. EPR studies of type 1 enzymes exhibit a distinctive signal with small hyperfine coupling constants ($\leq 9 \times 10^{-3} cm^{-1}$). This hyperfine coupling results from the interaction of Cu^{2+} unpaired electron with the copper spin ($I = 3/2$).

Type 2: The copper center is responsible for the oxygenation of substrates. The main examples of this class are oxidase reductase and the Cu-Zn superoxide dismutase enzyme. They show weaker absorption bands with higher EPR hyperfine coupling constants (Table 7.5). The most adoptable geometries for these types are trigonal, square planar, and tetragonally distorted octahedral, depending on the coordination number (Figure 7.2) [1,2].

Type 3: Copper enzymes contain binuclear Cu^{2+} (d^9) associated with antiferromagnetic coupling, making them EPR silent (Table 7.5). These enzymes exhibit a strong absorption band at 345 nm, caused by ligand-to-metal charge transfer. In this case, both copper atoms are five-coordinated with trigonal bipyramidal or square pyramidal stereo-orientation, depending on the nature of the protein. The most common example of a type 3 center is hemocyanin, which is accountable for oxygen transport. The attachment of a dioxygen molecule to the copper center occurs via a sidewise approach, resulting in $\eta^2{:}\eta^2$ per-oxo confirmation (Figure 7.2).

CLASSIFICATION OF COPPER ENZYMES

Type 1

Azurin

Azurin is a small metalloenzyme consisting of 128 amino acid residues. These enzymes are involved in electron transfer reactions in photosynthesis and respiration processes of bacterial systems. These blue proteins, as mentioned above, show an absorption band at 600 nm with an extinction coefficient greater than

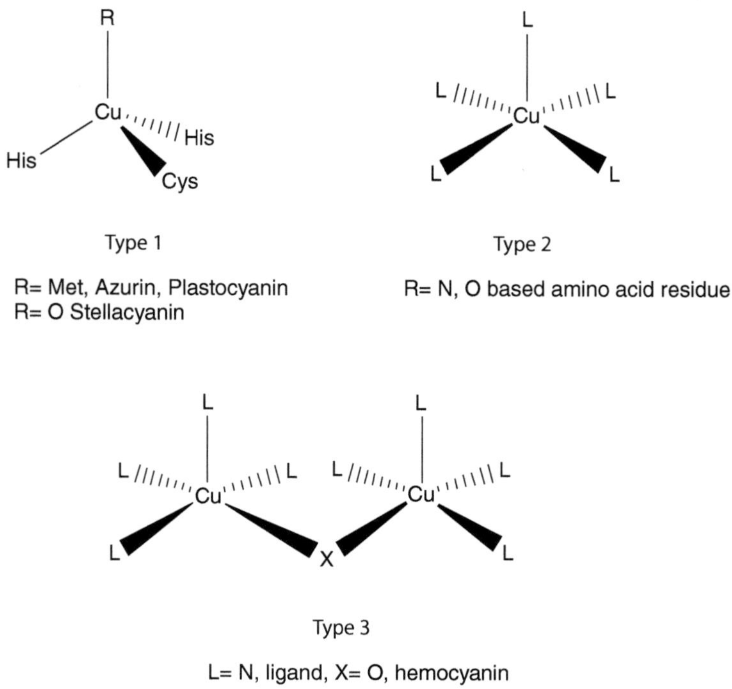

FIGURE 7.2 Geometrical aspects of type 1–3 copper enzymes.

3,000 M/cm. The appearance of a dark blue color has attracted further research to explore the geometry of the metal site in the protein. Strikingly, EPR studies displayed an intensive property, with high g values and low A values. This outcome suggests greater delocalization of the unpaired electron compared to other copper (II) complexes. During electron transfer, there is a switching between Cu^{+1} to Cu^{2+}, as a consequence, the coordination environment around the copper center compromises between Cu^{+1} trigonal and Cu^{2+} trigonal bipyramidal geometry [1].

X-ray crystallographic analysis corresponds to the ligation of the thiolate sulfur of Cys 112 and the imidazole nitrogen of His 116 and His 117 in distorted arrangement, with an average bond distance of 2.08 Å. A conciliation of ligands between the soft cysteine for copper I and harder for histidine imidazole ligands for copper II center takes place in a rhythmic way (Table 7.6). The presence of the thioether of Met 121 and the carbonyl oxygen of Glu 145 occupies the axial position in the Cu(II) state. Baker et al. reported minimal structural change during oxidation and reduction forms [2–4]. It is easy to conclude that both structures favor rapid electron transfer with low organizational energy.

Plastocyanin

Plastocyanin acts as a bridge for the electron transfer process between photosystem I and II in plants. Similar to azurin, its tertiary structure consists of 97–105 amino acid residues. The absence of a disulfide bond differentiates it from azurin.

TABLE 7.6
Selected Bond Lengths of Azurina Denitrificaus

Azurin	Cu(I) Bond Lengths (Å)	Cu(II) Bond Lengths (Å)
Cu-N (His 46)	2.13	2.08
Cu-N (His 117)	2.05	2.0
Cu-S (Cys 112)	2.26	2.15
Cu-S (Met 121)	3.23	3.11
Cu-O (Gly 45)	3.22	3.13

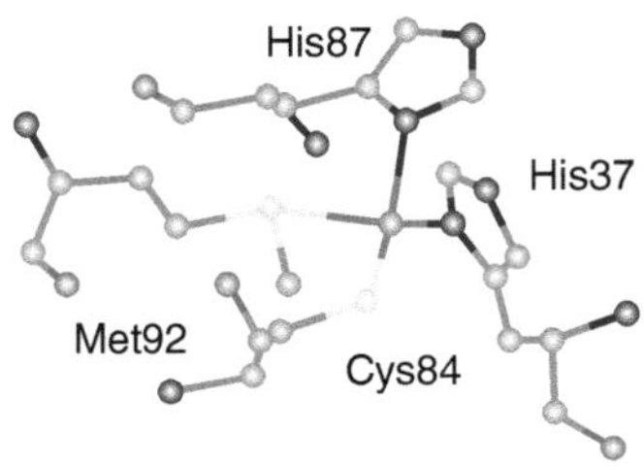

FIGURE 7.3 Geometrical orientation of poplar plastocyanin. (Reused with permission from Ref [5].)

TABLE 7.7
Selected Bond Lengths of Poplar Plastocynin

Plastocynin	Cu(I) Bond Lengths Å	Cu(II) Bond Lengths Å
Cu-S (Cys 84)	2.17	2.13
Cu-S (Met 92)	2.87	2.90
Cu-N (His 37)	2.13	2.04
Cu-N (His 87)	2.39	2.10

X-ray crystallography indicates a tetrahedral arrangement of ligands His 37, His 84, His 87, and Met 92 around the copper metal (Figure 7.3 and Table 7.7). The presence of protein surface exposed of copper site which is 6 Å deeper inside the hydrophobic core. Plastocyanin enzymes contain two potential binding sites for the physiological redox center, i.e., the north and east sites. On the north side, a hydrophobic patch is 6 Å away from the metal, whereas on the east side, an acidic patch is 15 Å away from the copper [1,4]. This acidic patch provides providing binding sites for electron proteins, cofactor, or substrates of a positively charged nature. The hydrophobic covering is expected to create a pocket in which the copper ion lies.

TYPE 2

Superoxide Dismutase

Superoxide dismutase metalloenzymes function to disproportionate the reactive superoxide ion/radical to hydrogen peroxide (Equation 7.1). However, the hydrogen peroxide itself is a harmful moiety for organisms. Therefore, it is scavenged from the body in a disproportionate manner to O_2 and H_2O, either by the catalase metalloenzyme or the peroxidase enzyme. In addition to Cu-Zn dismutase, other SOD forms have also been structurally characterized, in which the redox-active center is occupied by manganese (mitochondria and bacteria), iron (bacteria and plants), and nickel metals. Here, our main priority is to explore Cu-Zn SOD [5,6], which is also referred to as 'brass enzyme' based on its heterometallic core.

$$2O_2^- + 2H^+ \rightleftharpoons H_2O_2 + O_2 \tag{7.1}$$

Cu-Zn Superoxide Dismutase

Copper-containing SOD has been isolated from erythrocytes and termed erythrocuprin. Similarly, isolation from the liver is called hepatocuprin, and so on. As mentioned above, these are capable of catalyzing the disproportionation of the superoxide ion [6]. A simple mechanistic study indicates that two diffusion-controlled steps are involved in this process. One involves the reduction of Cu(II) to Cu(I) by the superoxide ion, and the other is based on the oxidation of metal center (Cu(I) to Cu(II)) by another mole of the superoxide moiety in the presence of a proton (Equation 7.2).

$$\begin{aligned} &SODCu^{2+} + O_2^{\bullet -} \rightarrow SODcu^{+} + O_2 \\ &SODCu^{+} + O_2^{\bullet -} + 2H^{+} \rightarrow SODcu^{2} + H_2O_2 \end{aligned} \tag{7.2}$$

These reactions are based on 'Pull- push/Ping-pong' mechanism because of the shuttling between common oxidation states of copper in a back-and-forth fashion. The protonation reaction in the second step makes both reactions thermodynamically feasible.

Structure

Cu-Zn SOD enzymes consist of two identical subunits held together by hydrophobic interactions. Each subunit is oriented as a flat cylindrical barrel, made of an antiparallel chain attached to an external loop. This protein structure is known as a β-barrel fold. These subunits contain one Cu and one Zn atom, with a molecular weight of 2×16 kDa. The imidazole ring of histidine 61 acts as a bridge between both metal centers, holding them 6.3 Å apart (Figure 7.4). The presence of a hydrogen bond between Arg 141 and the carbonyl of Cys 55, as well as adisulfide bond between Cys 55 and Cys 144, assists in maintaining the tertiary structure of the

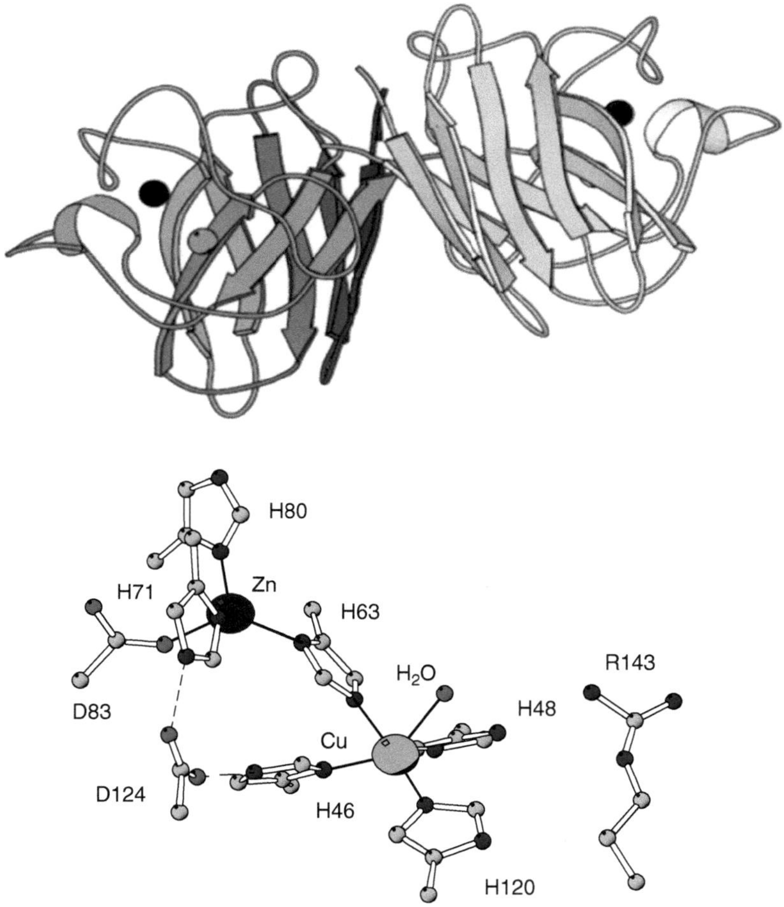

FIGURE 7.4 Elucidation of Human SOD β strand and its crystal structure. (Reused with permission from Ref. [6].)

protein [6,7]. The coordination core of Cu(II) is defined by the ligation of the nitrogen atoms of His 44, his 46, His 118, and the bridged His 61 (Figure 7.4). It is also believed to include water as a fifth ligand. The orientation of His 44 and His 46 is *trans*-configured in a distorted square planar CuN_4 coordination core. The tripeptide His44-Val45-His46 blocks access to copper ions from one side of the CuN_4 plane. However, the other side is free to access the solvent through a conical channel, 4 Å wide at the copper, which opens to the solution at the protein surface.

In addition to the copper active site, the floor of the cavity also contains zinc ions. The geometry of Zn(II) is tetrahedrally coordinated with histidine residues His 61, His 69, His 78, while the remaining fourth coordination site is ligated to the carboxylate group of Asp 81 (Figure 7.4). Till date, imidazolate-bridged bimetallic Cu_2Zn_2SOD is a novel example in enzymes as well as in the coordination chemistry of its own type, exhibiting unique bridging motifs.

Role of Zinc in SOD

The role of zinc (II) ion in Cu_2Zn_2SOD is ambiguous in nature. General considerations about its multifunctional behavior are discussed as follows:

1. Zn(II) ions play an important role in generating an electric field gradient that attracts superoxide ions into the conical cavity. Since the product of the disproportionation of O_2^- results in the formation of neutral products O_2 and H_2O_2, which can diffuse out, but O_2^- cannot.
2. Another possible role of zinc is to provide thermal stability to the metalloenzyme. It has been observed that the spectral properties of the Cu_2Zn_2SOD enzyme remain unchanged even after a 7-minute incubation in a 10 mM buffer at 75°C.

Mechanism

To understand the mechanistic perspective, a catalytic cycle has been proposed in Scheme 7.1. An important aspect of the mechanism is the breaking and reforming of bridged imidazolate rings during the catalytic cycle. A noteworthy fact is that the rate at which this cleavage occurs should be faster than the turnover rate of the enzymes [7]. Details are provided below:

1. Superoxide is electrostatically guided into the active site channel and binds to the Cu(II) center. Subsequently, it stabilizes with Arg 141 through hydrogen bonding, displacing water molecules.
2. Electron transfer takes place from the superoxide ion to the Cu(II) metal center via an inner-sphere electron transfer mechanism. This transfer results in the bond cleavage of Cu-His61 (bridged) along with the protonation of the imidazole nitrogen of the bridged His 61. As a result, a neutral oxygen molecule diffuses out of the active site channel.
3. In the second half of the catalytic cycle, a second molecule of superoxide enters the active site channel and associates with Arg 143 and water molecules through hydrogen bonding.
4. Finally, electrons are accepted from Cu(I) to superoxide via an outer-sphere electron transfer. Simultaneously, superoxide subtracts protons from H-bonded water molecules as well as imidazole-attached protons of the bridging His 61. As a consequence, the Cu (II) ion is reformed and re-bonded with His 61. Electrically neutral H_2O_2 diffuses out into the active site channel.

Spectroscopic Evidence

In support of the above mechanism, EXAFS studies have provided fruitful outcomes. They indicate the existence of trigonal planar geometry around the copper center, ligated with three histidine molecules, and the elongation of Cu-N (bridged His 61) to 3.2 Å during reduction (Table 7.8). However, square pyramidal geometry has been observed with the coordination of three histidine, water, and

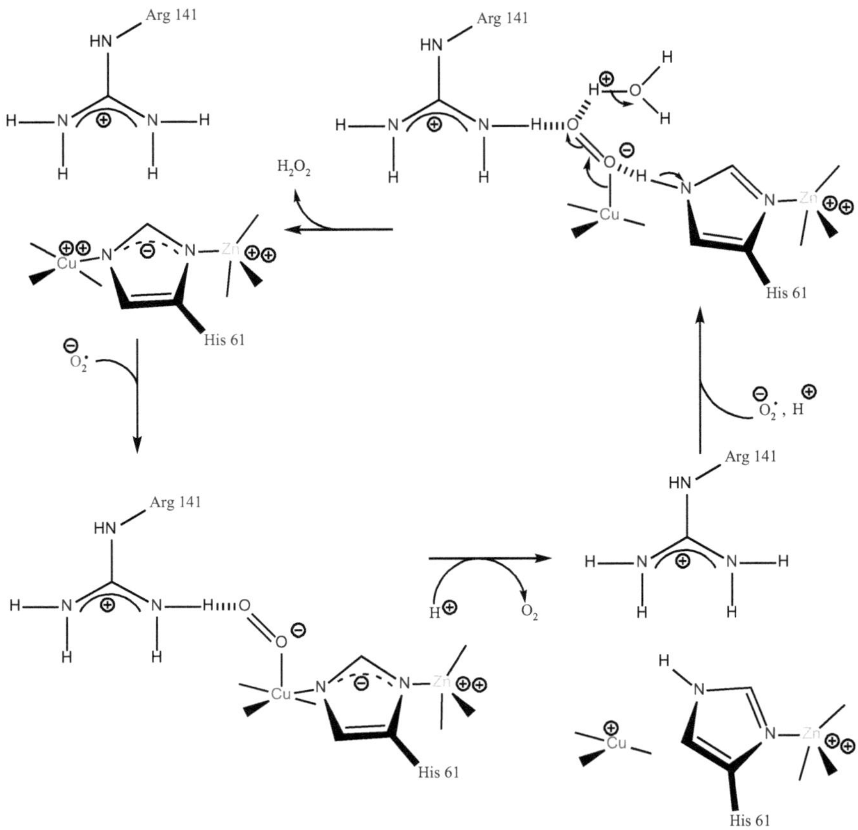

SCHEME 7.1 Catalytic mechanism of Cu, Zn superoxide dismutase. (Redrawn from Ref. [7].)

TABLE 7.8
Comparison of Oxidized and Reduced Form of Cu, Zn Superoxide Dismutase

S.N.	Property	Reduced Form	Oxidized Form
1	Coordination number	3	5
2	Geometry	Trigonal planar	Square pyramidal
3	Ligand	3 histidine, No bridging His 61 and H_2O	His 61 and H_2O act as 4th and 5th axial ligand
4	Cu-Zn interaction	6.6Å	6.1Å
5	Cu-N his 61	3.2Å	2.2Å

the bridged His 61 residue, with the shortening of Cu-N (His 61) to a bond distance of 2.2 Å in the oxidized form (Figure 7.5). During the oxidation and reduction process, no change has been noticed in the in Zn-N (His 61) bond length.

FIGURE 7.5 Depiction of changes occurring in the Cu-Zn sphere of Cu, Zn superoxide dismutase.

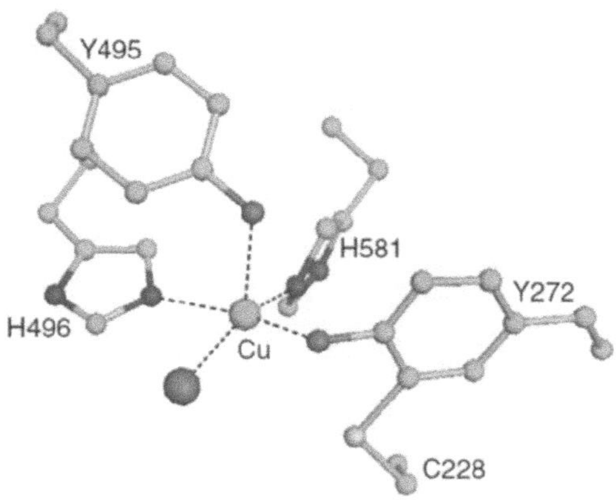

FIGURE 7.6 Ribbon model of three domains of galactose oxidase. (Reused with permission Ref. [8].)

Galactose Oxidase

Galactose oxidase is a type 2 copper protein (non-blue oxidase) that consists of a single polypeptide chain of 639 amino acids with a molecular weight of 68 kDa. Galactose oxidase catalyzes the oxidation of a primary alcohol to an aldehyde (Equation 7.3), and reduces oxygen to H_2O_2 via a two-electron reduction [2,4,9]:

$$RCH_2OH + O_2 \rightarrow RCHO + H_2O \qquad (7.3)$$

Galactose oxidase has low substrate specificity but absolute stereospecificity. It catalyzes a wide range of compounds, from small molecules like propane-1, 2-diol to polysaccharides.

Structure

The three-dimensional structure of galactose oxidase consists entirely of β strands with a small part of α helix (Figure 7.6). The enzyme can be divided into three domains: domain 1 (1–155 residue), domain 2 (156–532 residue), and domain 3 (533–639 residues).

Domain 1 contains the first 8β strands as well as binding sites for sodium ions and D-galactose. It is located a short distance from the copper center. Domain 1 acts as a 'chaperone' for the folding of the enzyme (Figure 7.6). The D-galactose binding site of domain 1 may possibly require attaching the enzyme to the cell wall of trees, the natural habitat of Dactylium dendroides. The function of the sodium ion is still unknown.

Domain 2 is the largest of the three domains, having 28 β strands oriented in a seven-fold geometry and a short α helix. It also contains 3–4 protein ligands, a copper ion, along with two acetate ions. The folding of domain 2 develops a large cavity in the middle to accommodate the β strands of domain 3 (Figure 7.6).

Domain 3 is located to the side of domain 2, opposite the copper center. It consists of seven β strands, two of which extend into the cavity of domain 2. The primary function of this domain is to provide structural stabilization (Figure 7.6).

Copper Center

Galactose oxidase catalyzes a two-electron reduction process. In this case, the divalent copper ion Cu^{2+} is antiferromagnetically coupled with a tyrosyl radical to make copper center EPR silent. The central metal ion is surrounded by several aromatic side chains capable of generating or stabilizing free radicals. The geometry around the copper ion is square pyramidal, with four ligands, Tyr 272, His 496, 581, and acetate ion, defining the plane (Figure 7.7) [10,11]. However, Tyr 495 act as the fifth ligand at axial position, located farther from the central ion (Table 7.9).

Substrate Binding

The copper center exists in a pocket of domain 2 along with the acetate ion and water. This pocket is structurally complementary to the chair conformation of D-galactose. The O-6 atom of galactose oxidase substitutes the acetate ion and water. As a result, it directly bonds to the copper ion. The presence of steric

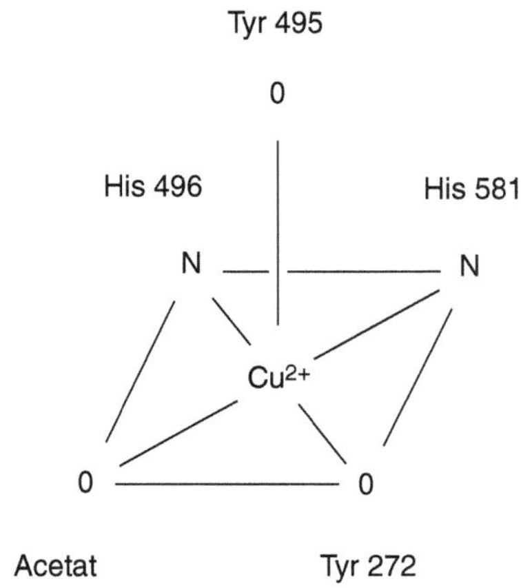

FIGURE 7.7 Model of type 2 copper center of galactose oxidase.

TABLE 7.9
Selected Bond Length of Galactose Oxidase Extracted from Dactylium dendroides

Galactose Oxidase	Bond Length at pH = 4.5 (Å)	Bond Length at pH = 7.0 (Å)
Cu-O tyr 272	1.9	1.9
Cu-O his 496	2.1	2.2
Cu-O his 581	2.2	2.2
Cu-O tyr 495	2.7	2.6
Cu-O acetate	2.3	–
Cu-O water	–	2.8

hindrance opposes the binding of L-galactose to the active center. Similar to other nonblue oxidases, galactose oxidase avoids the use of an external cofactor for the catalytic process by developing a modification of a residue of its own polypeptide chain. The amino acid residue that is modified is tyrosine. This modified tyrosine is a site that carries a free radical.

Type 3

Hemocyanin

Hemocyanins are multimeric proteins containing several subunits, depending on the extraction source. Molluscan hemocyanins have 20 subunits with a molecular weight of 200 kDa. However, arthropods have six asymmetric units containing dinuclear copper centers per unit (a total of 48 dinuclear copper present in quaternary structure), resulting in a molecular weight of 460 kDa. This extracellular metalloprotein acts as an oxygen carrier by the ligation of oxygen molecules to copper sites. The presence of Cu(II) ions in oxy-hemocyanin results 'blue blood' in arthropods and mollusks [1,10,11].

Deoxy-hemocyanin contains a dicuprous core coordinated with three histidine residues on each copper center (Table 7.10). The two copper metals are

TABLE 7.10
Selected Bond Lengths of Limulus Polyphemus Hemocyanin

Cu_A		Cu_B	
Bond	Bond Length Å	Bond	Bond Length Å
Cu_A-Cu_B	4.61		
Cu_A-N His173	2.10	Cu_B-N His324	2.16
Cu_A-N His177	2.01	Cu_B-N His328	2.08
Cu_A-N His 204	1.04	Cu_B-N His364	1.92

separated by an internuclear distance of 3.7 Å. This separation creates an empty cavity between both the metals, allowing the accumulation of a dioxygen molecule (Figure 7.8). Spectroscopic and magnetic studies conclude the presence of strong antiferromagnetic coupling between the two Cu (II) d^9 ions, making the binuclear center diamagnetic. Oxy-hemocyanin (Hc) exhibits an intense absorption band at 340 nm and a weaker one at 580 nm. These transitions are assigned to ligand-to-metal charge transfer bands. The resonance Raman spectrum of oxy-Hc displays a single band at 745–750 cm^{-1} for O-O stretching vibrations, supporting the existence of 1-μ-peroxo dicopper (II) center [1,2,11]. However, two coordination modes for binding oxygen to the binuclear sites are possible. One arrangement corresponds to end-on attachment of oxygen (Figure 7.9, structure A), whereas the second suggests side-on bonding of O_2 to the same site (Figure 7.9, structure B).

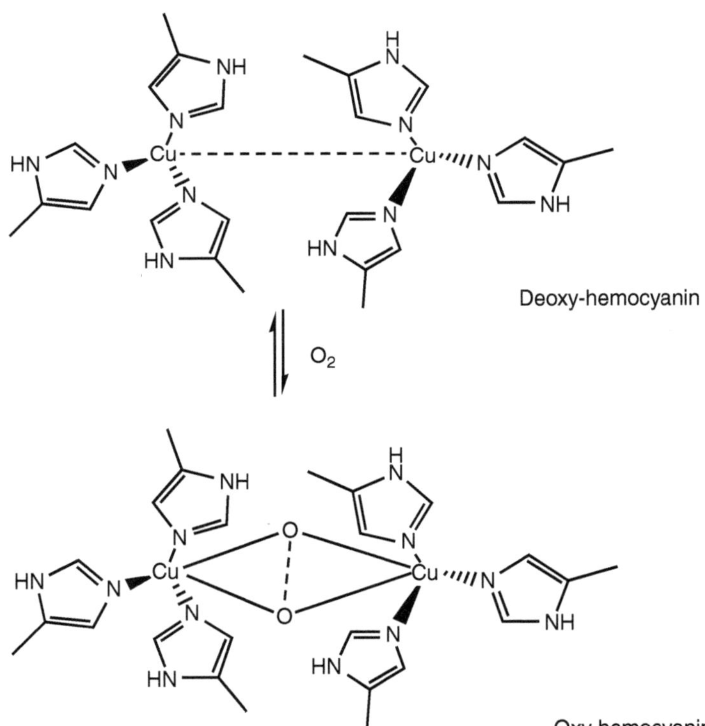

FIGURE 7.8 Configurational changes of hemocyanin during oxygenation.

FIGURE 7.9 Plausible arrangement of dioxygen to copper metalloenzyme.

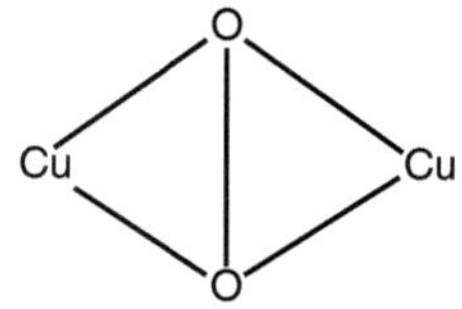

FIGURE 7.10 Attachment of oxygen molecule in η^2:η^2 fashion.

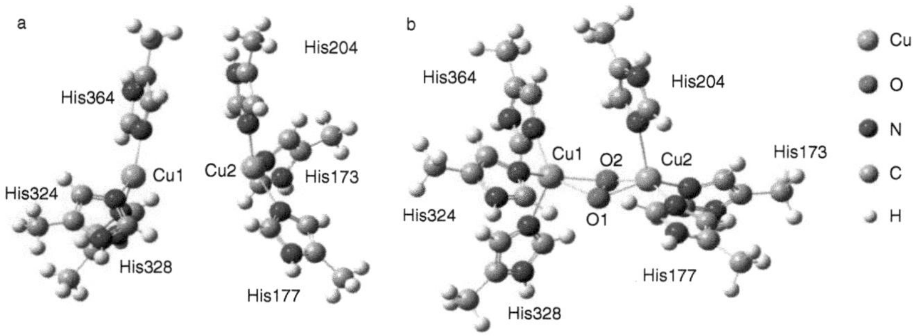

FIGURE 7.11 Structure of de-oxyhemocyanin with a binuclear copper core. (Reused with permission from Ref. [12].)

The former binding mode results in two O–O stretching bonds; however, the latter has a single frequency. We observed one resonance in the Raman spectrum. The more likely binding mode supports side-on bonding. Recently, a model compound showed similar spectroscopic properties to oxy-hemocyanin, in which two copper centers are attached by η^2:η^2 peroxide mode [2,11], as illustrated in Figure 7.10.

Simultaneously, the crystal structure of hemocyanin extracted from horseshoe crab also supported the presence of η^2:η^2 peroxide moiety (Figure 7.11).

Tyrosinase

The group of mono-oxygenase contains enzymes such as cytochrome P-450, domain-β-mono-oxygenase, and tyrosinase proteins present in almost all types of organisms, from bacteria to mammals [10,11]. Oxidation of tyrosine leads to L-dopa, which, in turn, by oxidation and polymerization, results in melanin formation and other related pigments, developing a range from red to dark brown (Scheme 7.2). It catalyzes the conversion of monophenols to o-diphenols (cresolase activity) and two-electron oxidations of o-diphenols to quinones.

The structure of tyrosinase is very similar to that of hemocyanin (Table 7.11) in which the binuclear center is ligated with six histidine residues, vacating one position on each copper to coordinate with oxygen. This oxygen-binding site is accessible to the solvent and substrate for their final conversion to di-quinone. The functioning of the Cu-enzyme to catalyze the incorporation of oxygen into the substrate, resulting in the oxygenation of monophenol, is discussed in Scheme 7.3.

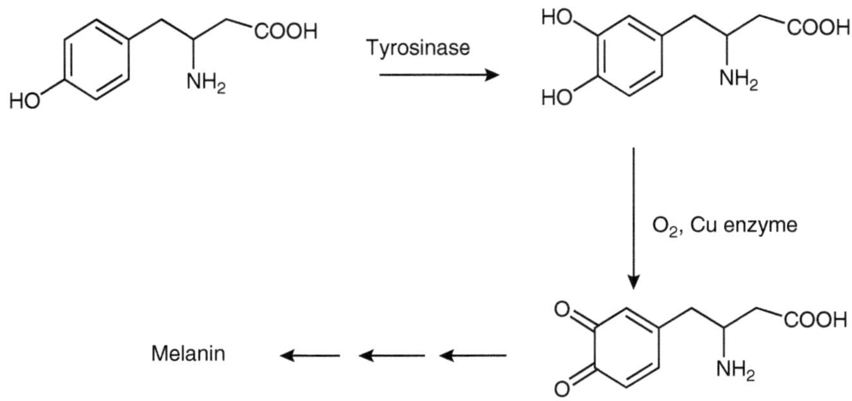

SCHEME 7.2 Oxidation of mono-phenol to melanin by applying tyrosinase enzyme. (Redrawn from Ref [6].)

TABLE 7.11
Comparative Studies of Various Type Copper Metalloenzymes and Their Function

Enzyme	Oxidation State and Geometry	Function
Azurin	Cu^{2+}(TBP): Cys, 2 His (in plane) Met., and O atom of gly (axial) Cu^{1+} (trigonal): Cys, 2 His	$Azu_{(ox)} + e^- \rightarrow Azu_{(red)}$
Plastocyanin	Cu^{2+} (Td): 2 His, Cys, Met. Cu^{1+} (Td): 2 His, Cys, Met.	$Pc_{(ox)} + e^- \leftrightarrow Pc_{(red)}$
Ascorbate oxidase	Cu^{2+} (Td): 2 His, Cys, Met. Cu^{2+} (trigonal): 2 His, OH^- Cu^{2+} (trigonal): 3 His	L-ascorbase + O_2 ↓ Dehydro-ascorbate
Tyrosinase	Cu^{2+} (Sq. pyr.): 3 His and side on η^2: η^2 peroxy O_2^{2-} Cu^{1+} (trigonal): 3 His	Monophenol + O_2 + $2H^+$ ↓ o-diphenol + H_2O
Hemocyanin	Cu^{2+} (Sq. pyr.): 3 His and side on η^2: η^2 peroxy O_2^{2-} Cu^{1+} (trigonal): 3 His	$Hc + O_2 \rightarrow Hc.O_2$ transport of hemocyanin

1. Deoxy tyrosinase is reversibly coordinated to an oxygen molecule, featuring a μ:η^2:η^2 peroxide structure of oxy-hemocyanin.
2. Oxy-hemocyanin enables the orientation of the phenolate ion (substrate) through π-π interaction directed by His 194 at Cu_B.

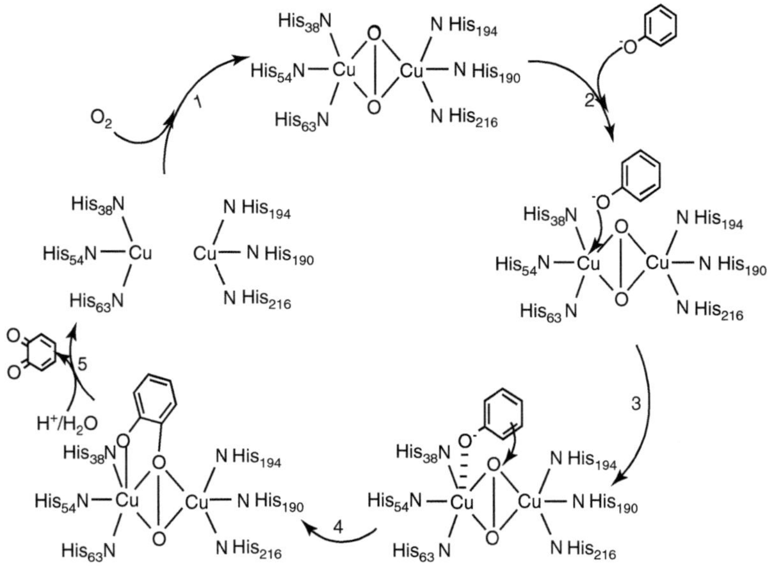

SCHEME 7.3 Catalytic cycle of tyrosinase enzyme (monophenol to quinone).

3. The phenolate oxygen is attracted toward the Cu_A center and result in a bond with oxygen. This resulting coordination indicates the movement of the O–O peroxy linkage from a horizontal to a vertical direction.
4. At this stage, chelated complexes are formed by the electrophilic attack of the O–O moiety on the aromatic ring, which results in the bond breaking of the O–O bond by transferring its electron density to the antibonding σ* orbital. The coordination of catechol features a $\mu:\eta^2:\eta^1$ binding mode.
5. Protonation of the opposite hydroxo ligand gives dissociable water, which in turn leads to the oxidized o-quinone structure by intramolecular electron transfer.

CONCLUSIONS

In summary, an attempt has been made to discuss the evolution of copper-containing enzymes in three different ways. A critical analysis related to their functioning, as well as mechanistic prospects, has been explained to correlate phylogenetic origins, i.e., blue proteins, metallothionines, and type 2 copper proteins, which are present in single-celled organisms. Hemocyanin, galactose oxidase, and others whose functions are correlated to non-copper proteins exist in multicellular life. The formulation of mono- and oligonuclear copper enzymes in distorted tetrahedral arrangements assists in catalyzing various important reactions such as oxygenation, electron transfer, and photosynthesis.

REFERENCES

1. S. J. Lippard, J. M. Berg, *Principles of Bioinorganic Chemistry*, University Science Books, Mill Valley, CA, 1994.
2. R. M. Roat-Malone, *Bioinorganic Chemistry: A Short Course*, John Wiley & Sons, Inc., New Jersey, 2002, Pages 1–365.
3. R. W. Hay, *Bio-Inorganic Chemistry*, Ellis Horwood Limited, Halsted Press, New York, 1984.
4. W. Kaim, B. Schwederski, A. Klein, *Bioinorganic Chemistry: Inorganic Elements in the Chemistry of Life*, John Wiley & Sons, Inc., New Jersey, 2013, Pages 1–426.
5. Y. Lu, From biology to nanotechnology. In J.A. McCleverty, E. Constable, T.J. Meyer (Eds.), *Comprehensive Coordination Chemistry II* (Volume 8), Elsevier, Amsterdam, the Netherland, 2003, Pages 91–122.
6. (a) A. F. Miller, Superoxide dismutases: active sites that save, but a protein that kills, *Curr. Opin. Chem. Biol.* 2004, 8, 162–168. (b) P. Agarwal, M. Singh, J. Singh, R. P. Singh, Microbial tyrosinases: a novel enzyme, structural features, and applications. In P. Shukla (Ed.), *Applied Microbiology and Bioengineering*, Academic Press, New York, 2019, Pages 3–19.
7. E. L. Solomon, M. D. Lowery, Electronic structure contributions to function in bioinorganic chemistry, Science, 1993, 259, 1575–1581.
8. A. Messerschmidt, Copper metalloenzymes. In H.-W. Liu, L. Mander (Eds.), *Comprehensive Natural Product* (Volume 8), Elsevier Ltd, Kidlington, 2010, Pages 489–545.
9. R. J. P. Williams, Metal ions in biological systems, *Biol. Rev.* 1953, 28, 1–412.
10. K.D. Karlin, Z. Tyekl, *Bioinorganic Chemistry of Copper*, Chapman and Hall, New York, 1993, Pages 1–501.
11. I. Bertini, H. B. Gray, S. Lippard, J. Valentine, *Bioinorganic Chemistry*, University Science Books, Mill Valley, CA, 1994, Pages 1–593.
12. T. Saito, Y. Kitagawa, M. Shoji, Y. Nakanishi, M. Ito, T. Kawakami, K. Yamaguchi, Theoretical studies on the structure and effective exchange integral (Jab) of an active site in oxyhemocyanin (oxyHc) by using approximately spin-projected geometry optimization (AP-opt) method, *Chem. Phy. Lett.* 2008, 456, 76–79.

8 Biochemistry of Zinc

INTRODUCTION

Zinc is an essential trace element for multicellular eukaryotes. An approximate amount of 3 g is required for the adult human body (70 kg body weight), which is second to iron in abundance when considering aggregates of trace elements. Most of the zinc (>95%) in our body is intracellular, with about 40% present in the nucleus, and the remaining 50% is found in cytoplasm, digestive system, and cell membrane [1,2]. The extracellular content is associated with albumin and keratin in the digestive system. Its vulnerable presence makes it an obvious consideration that zinc deficiency encompasses several severe effects, such as reduced sense of taste, lack of appetite, growth disorders, painful movements, loss of taste, and weakening of the immune system. High concentrations of zinc are present in the tissues of growing fetuses, infants, and seminal fluid, indicating the essentiality of the metal in various metabolic processes [3,4].

Physiological conditions favor the existence of zinc in its di-cationic state (2^{+}) in our biological system. This Zn(II) is diamagnetic, colorless, and has poor electronic excitation due to fully filled d-shell. The latter property, which brings zinc-related enzymes into the spotlight later compared to copper and iron, is mainly attributable to its spectroscopic silence. These limitations have been compensated for the development of various sensitive analytical techniques, such as atomic absorption spectroscopy (for zinc detection), NMR, and X-ray diffraction (structural confirmation). These spectroscopic developments have led to identifying more than 300 zinc-containing proteins so far. Most of these enzymes are involved in catalyzing hydrolysis processes of various metabolites [4]. Recent developments have shown their other functions related to structural considerations (oxidoreductases), stabilization (for insulin hormone), and regulation of transcriptase process. Many of these enzymes feature tetrahedral zinc coordination, occupied by three amino acid residues: histidine (N), glutamate (O), aspartate (O), and cysteine (S), with the remaining fourth active site being attached to water (Scheme 8.1).

The chronological discovery of zinc biology is divided into three phases (Figure 8.1). In the first phase, a remarkable conclusion about zinc's essentiality was made by Jules Raulin in 1869, who studied the dependency of Aspergillus on zinc for growth. The second phase considers the time elapsed until another groundbreaking discovery of a zinc enzyme, carbonic anhydrase (Scheme 8.2), which was first isolated in 1939 by Keilin and Mann, followed by the discovery of pancreatic bovine carboxypeptidase-A (1954) and number of zinc proteins in proteomes by Bretini et al. The current phase unfolds the

DOI: 10.1201/9781003432333-8

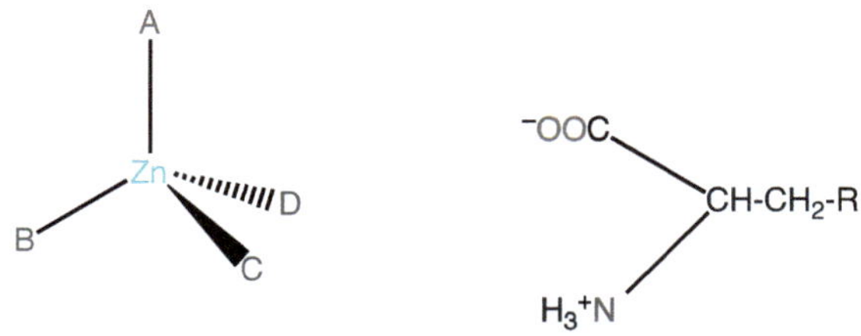

B, C, D = His, Glu, Asp, Cys
A = H_2O, amino acid

R	Amino acid
$-CH_2COOH$	Glutamate
Imidazole	Histidine
$-COOH$	Aspartate
$-SH$	Cysteine

SCHEME 8.1 Structural representation of zinc enzymes.

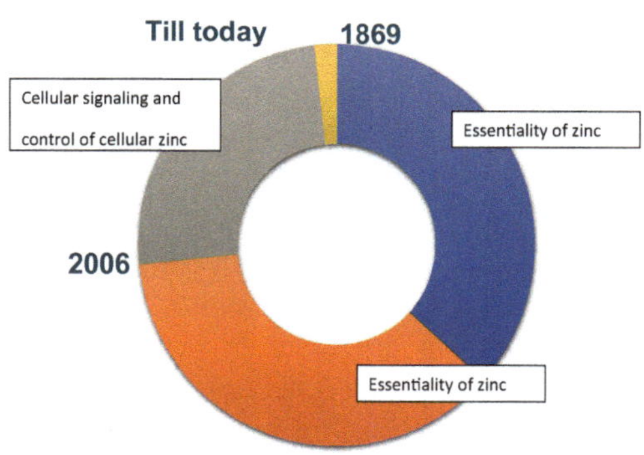

FIGURE 8.1 Depiction of various phases involved in zinc biology. The three phases of discoveries in zinc biology: Zinc as an essential nutrient, Zinc as a cofactor in proteins and proteomes, and discoveries related to the role of zinc that are open for future purposes.

role of Zn^{2+} in synaptic transmission (2006) [4]. This cellular signaling has not had a well-defined beginning, and many avenues remain open for further exploration.

A tremendous growth in zinc biology has occurred in recent years. The growing number of zinc dependent enzymes needs systematic segregation for a better understanding of their functioning in terms of catalytic, co-catalytic, and structural nature. In most zinc-based enzymes, the substrate interacts with protein residues inside the cavity and Zn^{2+} to activate bond-breaking and formation processes. A summary of zinc-based enzymes and their functions is mentioned in Table 8.1.

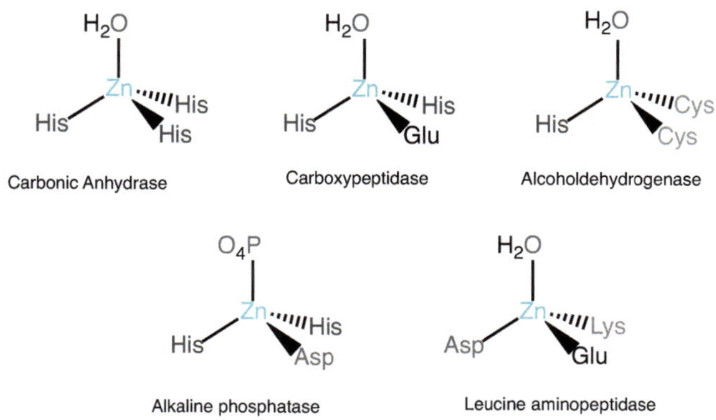

SCHEME 8.2 Classic examples of zinc-based enzymes.

TABLE 8.1
Several Zinc Enzymes and their Functions

Enzyme	Mol.wt. (kDa)	Ligands	Function
Carbonic anhydrase	30	3 His, $1H_2O$	Reversible hydrolysis of CO_2
Carboxypeptidase	34	2 His, 1 η^2 Glu, 1 H_2O	Hydrolysis of C-terminal peptide residue
Thermolysis	35	2 His, 1 η^2 Glu, 1 H_2O	Hydrolysis of peptides
Alcohol dehydrogenase	2×40	2 (2Cys, 1His, $1H_2O$)	Hydride transfer from alcohols via NAD+
Alkaline phosphatase	86	2His, 1 Asp	Hydrolysis of phosphodiesterase
TFIIIA	40	4 Cys, 4 His	Regulation of transcription process
Superoxide dismutase	2×16	2 (2His, 1-μ-His, 1Asp)	Disproportionation of O_2•–
Adenosine deaminase	140	3 His, 1 Asp	Purine metabolism (breakdown of adenosine from food)

RATIONAL CHARACTERISTIC OF ZINC EFFECTIVENESS FOR BIOLOGICAL SYSTEM

The characteristic features that make divalent zinc more adaptable for biological systems are as followed:

Lewis Acidity

The Lewis acidity of Zn(II) metal plays an important role in various biological processes without involving any electron transfer reactions. Normally, electron-deficient compounds are considered Lewis acid, such as H^+, carbocations, etc., which are prominently used as catalysts for condensation reactions (the reverse of hydrolysis reactions) [1]. An alternative to these catalysts is an electrophilic substrate with a higher effective nuclear charge (Lewis acid) for the activation of the substrate molecule. The rate of reaction can be accelerated through internal attacks within the coordination sphere of metal ions.

In zinc enzymes, zinc is always associated with a water molecule. It can rapidly exchange with the substrate due to the kinetically labile nature of zinc complexes. An available lone pair of oxygen atoms coordinated with the metal ion becomes formally positively charged, which in turn results in the easy release of a proton. Thus, presence of positive charge in metal lowers the pKa value of coordinated water compared to free. Thus, coordinated hydroxide is a strong nucleophile, weaker than free OH^- ion but better than water. The order of nucleophilicity is as follows:

$$H_3O^+ < H_2O-M^{2+} \sim H_2O < HO-M^+ \sim OH^-$$

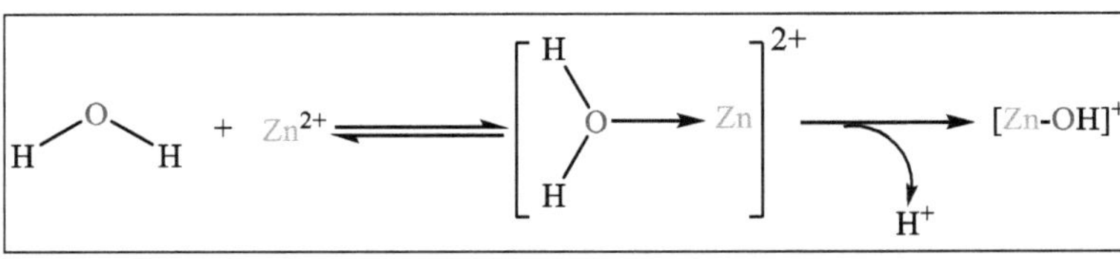

Hence, this property of the Zn^{2+} ion catalyzes the rapid attack on the nucleophilic substrate, e.g., H_2O, to yield $[Zn\text{-}OH]^+$. The formation of this species is a crucial step in enzymatic hydrolysis. In summary, the pK_a of coordinated water in zinc complex is governed by two factors: the coordination number and the total charge on the complex (Table 8.2). The pK_a value is directly related to the coordination number and inversely proportional to the charge. Therefore, greater charge on metal ion has greater attraction for the oxygen lone pair, thereby lowering pK_a.

Chemical Inertness

A fully filled electronic configuration, i.e., d^{10}, of Zinc (II) results in zero crystal field stabilization energy. Hence, the coordination number around the metal is mainly determined by the balance between bond energies and repulsion among

TABLE 8.2
pK_a Value of Various Zinc Enzymes

Enzyme	pKa (Value of Coordinated Water)
Carbonic anhydrase	6
Liver alcohol dehydrogenase	7.2
Coenzyme-free enzyme	9.2
NADH	11.2

TABLE 8.3
Average Zinc(II) Ligand Bond Distance in Coordination Complexes Having Four, Five, and Six-coordination Number

		Coordination Number		
S.N.	**Ligand**	**4**	**5**	**6**
1	H_2O	2.0	2.08	2.10
2	Pyridine	2.06	2.12	2.11
3	$RCOO^-$	1.95	2.02	2.07

the ligands. Among the possible coordination four, five and six, tetrahedral four-coordinated complexes are preferable due to shorter metal-ligand distances and less ligand repulsion compared to the other geometries. The repulsion can be both steric and electronic. In general, the zinc metal ion exists in a coordination number less than six. This adaptation provides an available binding site for the substrate in its coordination sphere (Table 8.3). Usually, the substrate binds to zinc either by substitution or by increasing the coordination number. This behavior supports its Lewis acidic behavior. In comparison to a proton, the available coordination site and the presence of a double positive charge favor electrophilicity due to the smaller charge density [1].

Structural Preferences

As discussed above, zinc complexes show an ease of interconversion between four- and five-coordinate numbers. The low energy barriers between these two coordination geometries are very important because the substrate may attach to the coordination sphere by substituting water or together with water. This rapid interconversion between four- and five-coordinate numbers increases the catalytic rate.

Conclusively, the good Lewis acidic nature, rapid conversion of the zinc metal ion to coordination numbers four, five, and six, and its preferential low coordination number contribute to the pKa of coordinated water, making aqua complex labile.

CARBONIC ANHYDRASE

Carbonic anhydrase enzymes are responsible for catalyzing the hydrolysis of carbon dioxide. This CO_2 hydration is crucial for important biological processes like photosynthesis (intake of CO_2) and respiration (CO_2 disposal) (Equation 8.1). These vital activities highlight the biological importance of CAs, which are present in animals, plants, and bacteria. At physiological pH, the uncatalyzed hydrolysis of CO_2 is a slow process, with a rate of 10^{-1} S^{-1} [2,3]. However, in the presence of an isoenzyme of CA, the rate of the reaction increases 10^7 times.

$$CO_2 + H_2O \rightarrow HCO_3^{\ -} + H^+ \quad (8.1)$$

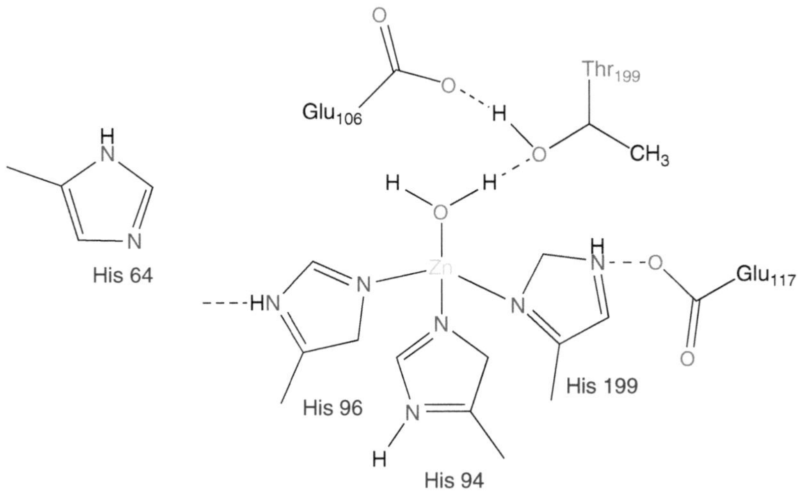

FIGURE 8.2 Structural orientation of carbonic anhydrase present in human RBCs, where zinc is in a tetrahedral environment coordinated by three histidine and one active site, i.e., water. It is an enzyme that converts carbon dioxide to bicarbonate ions in the lungs.

The structural conformation of carbonic anhydrase enzymes reflects that all isoenzymes (i.e., different enzymes from different sources catalyzing the same reaction, usually having homologous structures) have a shape like a rugby ball, consisting of a single polypeptide chain of 259 amino acids with mol.wt.30 kDa and an average size of $4\times4\times5.5\,nm^3$. The zinc ion is located at a 16 Å deep conical crevice, coordinated with three neutral histidines, i.e., His-94, His-96 and His-119, via their nitrogen atoms [1–4]. His-119 is associated with a glutamate residue through hydrogen bonding. It is believed that such interactions help establish the basicity of metal center. The fourth coordination site of Zn is occupied by water molecule (Figure 8.2). The presence of the H_2O molecule assists in forming the H-bond with the neighboring Thr-199, which in turn is H-bonded to Glu-106 [5].

MECHANISM

The mechanistic prospect (Scheme 8.3) involved in the reversible catalytic hydrolysis of carbon dioxide reactions is as follows:

1. Various studies show that a coordinated water molecule attached to the active site of carbonic anhydrase gets indirectly deprotonated and transfers the proton to His-64, resulting in more nucleophilic Zn-OH species.
2. This indirect transfer to His-64 passes the proton to the buffer molecule, which frees the histidine 64.
3. The hydroxide-bound species Zn^{II}-OH is nucleophilic enough to activate a symmetrical CO_2 molecule via several attacks from the SP^3 oxygen atom to the carbon (pull and push mechanism), resulting in a bicarbonate species.

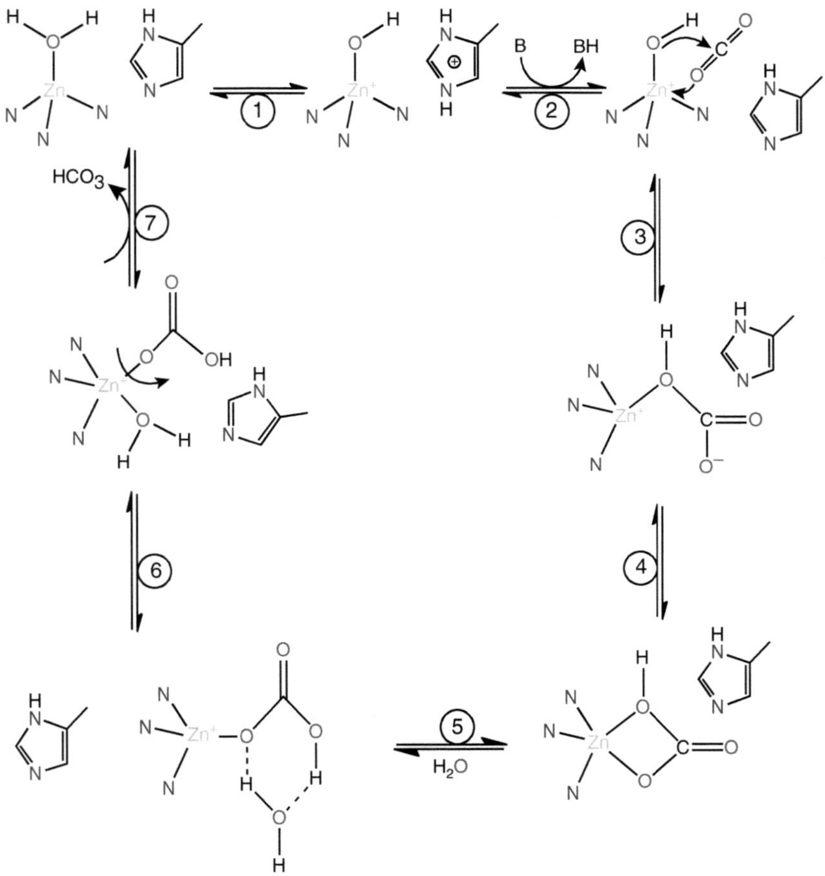

SCHEME 8.3 Mechanistic pathway of carbonic anhydrase enzyme.

4. The latter species converts to a cyclic intermediate, where the free oxygen atom interacts with zinc to transfer the proton to the distal oxygen center. With the help of a hydrogen-bonded water molecule, resulting in a six-membered cyclic species.
5. A five-coordinate species is obtained by the transformation of the cyclic species through the rupture of the H-bond in the presence of a higher concentration of dissolved CO_2, which provides a low barrier for the HCO_3^- detachment via an associative mechanism involving the coordination of a water molecule.

CARBOXY PEPTIDASE

An enzyme carboxypeptidase is responsible for cleaving the peptide bond, as per Equation 8.2. In this reaction, if R' is the C-terminal aromatic residue of the protein, it is termed carboxypeptidase-A; however, the presence of a basic group corresponds to carboxypeptidase-B (Equation 8.2). Both enzymes are synthesized in the pancreases and isolated from the same source.

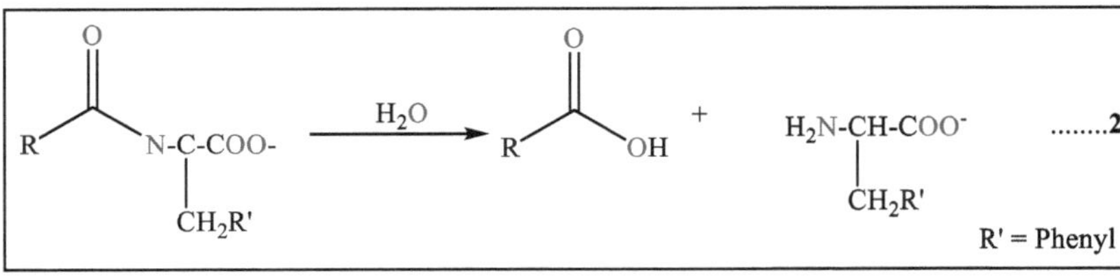

Carboxypeptidase A exists in three different forms: A_α, A_β, and Aγ, having 307, 305, and 300 amino acid residues, respectively (Figure 8.3). The high-resolution X-ray crystal structure of carboxypeptidase A_α depicts that the zinc atom is well embedded into the protein surface and coordinated by imidazole ring of His 69, as well as His 196 amino acid and bidentate carboxylate group of Glu 72 (Figure 8.4). The coordination of a water molecule to the active site of Zn (II) results in square pyramidal geometry. It has been noticed that the coordination

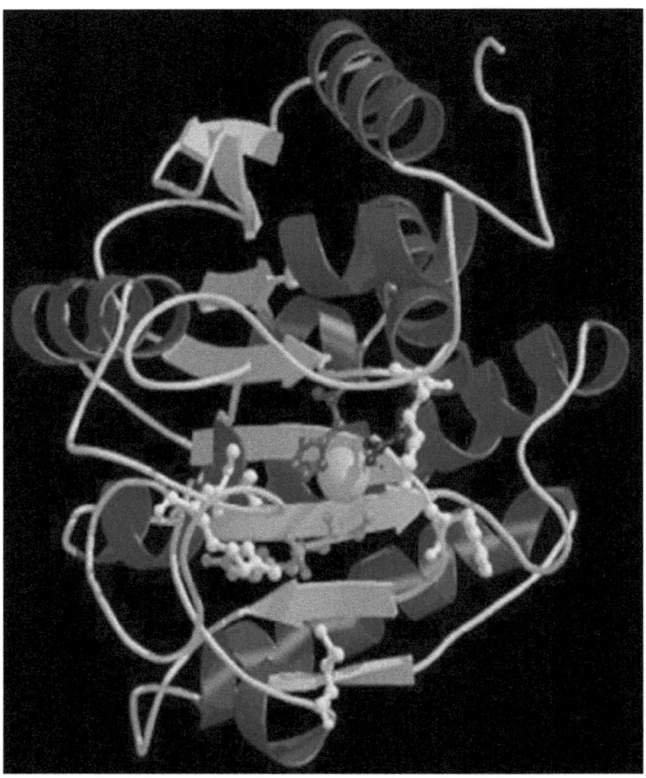

FIGURE 8.3 Richardson diagram of the cattle carboxypeptidase A/glycyl-L-tyrosine complex. (Reused with permission from Ref. [6]; Copyright (2023) Elsevier.)

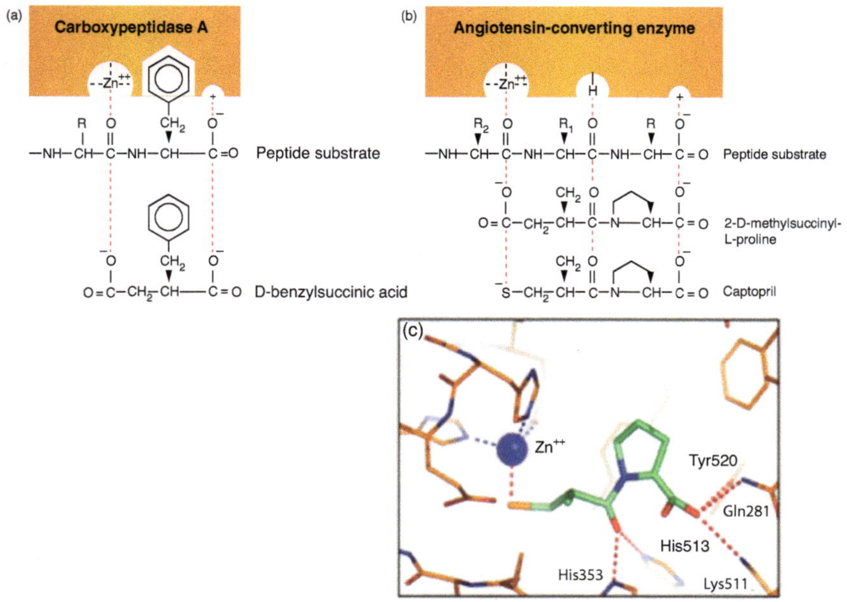

FIGURE 8.4 Schematic drawing of carboxypeptidase A, Carboxypeptidase A usually refers to the pancreatic exopeptidase, which contains a zinc (Zn^{2+}) metal center in a tetrahedral geometry with amino acid residues to facilitate catalysis and binding. (Reused with permission from Ref. [7]; Copyright (2023) Elsevier.)

of the substrate to the active site of the enzyme forces the bidentate carboxylate group of Glu 72 into a monodentate fashion. The ability of Glu or Asp residues to rearrange themselves in such a manner is termed a carboxylate shift [5,6]. Such inter-tuning assists in maintaining a constant coordination number even when various forms of substrate are bound to the metal. The hydrophobic pocket near the active site protrudes the aromatic ring at the C-terminal of the peptide chain. Hydrogen bonds exist between the coordinated water molecule and Glu 270. Various tyrosine and arginine residues positioned close to the active site assist in the process of substrate binding and activation.

STRUCTURE OF THERMOLYSIN

The structure of the endopeptidase thermolysin exhibits a close resemblance to carboxypeptidase. The striking similarities reveal the presence of three side chains, His 142, His 146, Glu 166, and water molecule coordinated to the zinc ion. The non-coordinating amino acids in both proteins are also similar. The Glu 143 and His 231 residues are positioned to analogously to Glu 270 and Arg 127, respectively in carboxypeptidase. These similar structural and mechanistic scenarios provide an example of convergent evolution, facilitating the catalysis of two related reactions [6,7].

MECHANISM

Although the peptide hydrolysis process is thermodynamically favorable, it is difficult to take place. It occurs either through a nucleophilic attack on the peptide carbonyl group to afford a highly reactive nucleophile or by activating the carbonyl group through polarization. These changes result in a tetrahedral intermediate that is obtained through nucleophilic attack at the carbonyl carbon. As a consequence, it helps to stabilize the amide nitrogen, making it a good leaving group by bond cleavage of the C-N bond. In order to understand the functioning of peptidase, two different mechanisms have been proposed. One school corresponds to the attack of the metal site on the carbonyl oxygen atom of the incoming peptide [7]. The second school suggests that hydroxide, coordinated to zinc metal, attacks the carbonyl carbon. Detailed mechanistic prospects involved in both mechanisms are discussed here:

MECHANISM A

1. Initiation step involves the coordination of the electrophilic metal center at carbonyl oxygen, which in turn activates the carbon for nucleophilic attack by Glutamate 270. This concerted attack on the carbonyl group forms an intermediate anhydride (Scheme 8.4).
2. Existence of a hydrogen bond between Glutamate and coordinated water facilitates bond cleavage to produce species B.

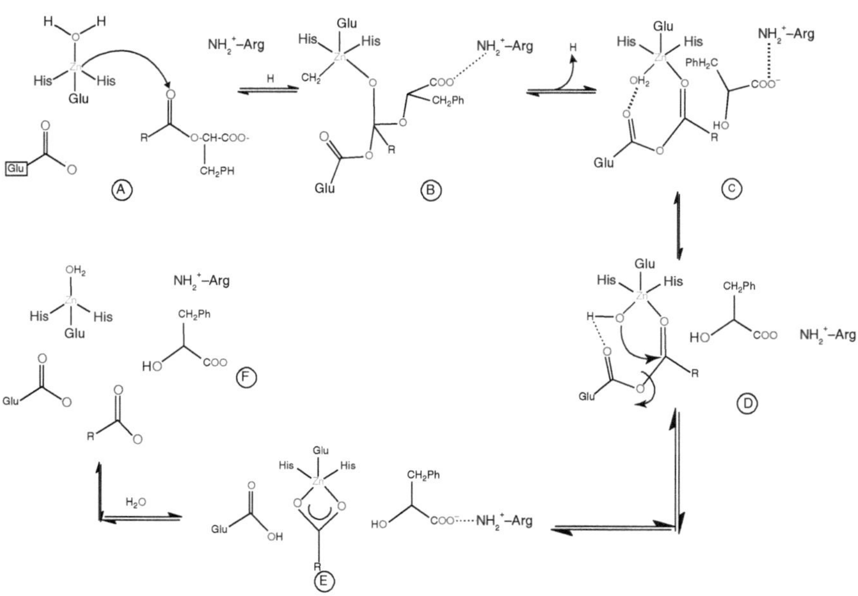

SCHEME 8.4 Mechanistic pathway of carboxypeptidase (pathway A).

3. Indirect deprotonation of the coordinated water attached to zinc facilitates the nucleophilic attack on the carbonyl carbon. As a consequence, the anhydride linkage is cleaved, and Glutamate is reformed.
4. Concerted step of bond forming and breaking (i.e., O-C and C=O, respectively) is confined to a zinc-mediated four-membered ring.
5. Hydrolysis of intermediate E leads to bond cleavage to obtain the expected product.

MECHANISM B

This hypothesis was widely accepted and proposed by Christiansen and Lipscomb, which showed a similar mechanistic pathway with respect to metal, similar to the carbonic anhydrase enzyme. The steps are as follow:

1. In this mechanism, the initiation step involves the interaction of the peptide chain with the arginine residue through hydrogen bond formation at the terminal carboxylate group. More stability is attained through the interaction of carbonyl oxygen of Arg 127 and NH group of Tyr 248. Consecutively, Glu 270 abstracts the proton from water molecule coordinated to zinc and metal hydroxide species.
2. At the second step, nucleophilic attack takes place on carbonyl carbon, which is polarized by Arg 127 through a hydrogen bond. Subsequently, an electrostatic interaction exists between the carbonyl oxygen and the metal center. Here, the five-coordination state of zinc is maintained by switching Glu72 from a bidentate to a monodentate fashion.
3. Finally, C-N bond cleavage takes place by the addition of a proton, coming from Glu 270, which turns into its ionized state. An additional proton comes from the coordinated carboxylic group to transform the $-NH_2$ group into the N^+H_3 moiety. In this way, the zinc metal moves back to regain its original state, with glutamate reclaimed to its bidentate manner, and the water molecule is added to the metal by cleaving the peptide bond (Scheme 8.5).

ZINC FINGERS

Zinc is an essential element required for the development of organisms. Higher accumulation of zinc in reproductive organs implies that it is not only important for the catalysis of various important reactions but also responsible for nucleic acid synthesis. In 1980, a special modular-shaped protein came into the picture that activates as well as regulates the transcription process. These units are termed zinc fingers (Zif) [8].

The first time zinc finger, i.e., TFIIIA, was isolated in 1980 from the ovaries of the African clawed toad (Xenopus). TFIIIA contains nine home domains in a particular sequence as mentioned in the scheme. These domains are highly

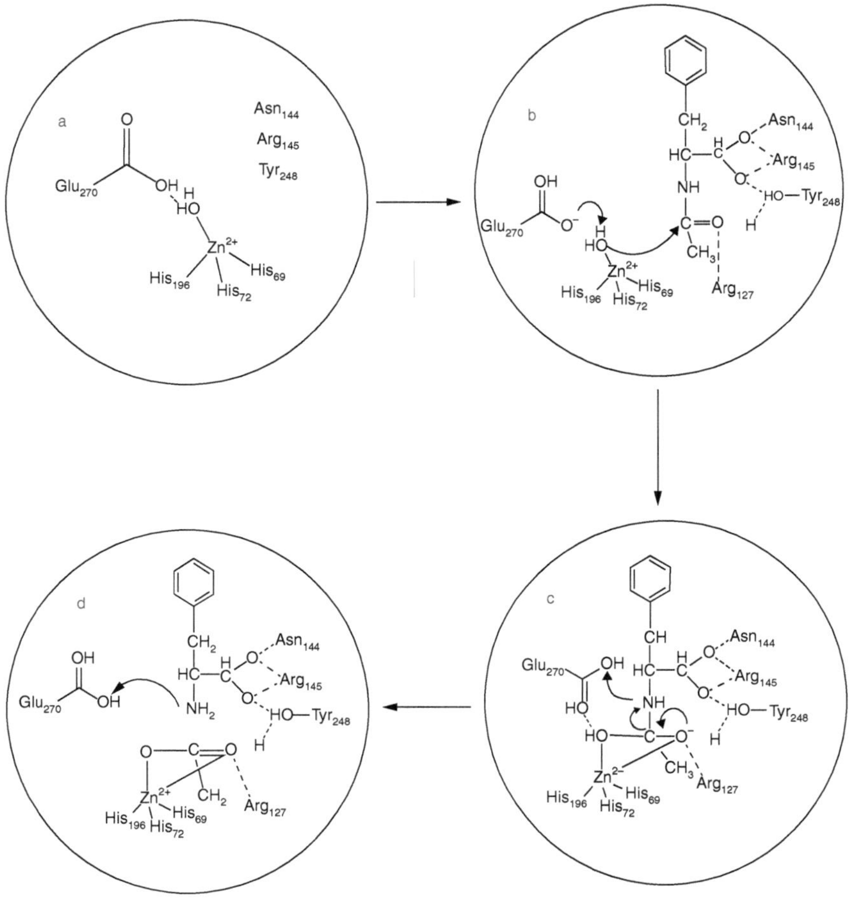

SCHEME 8.5 Mechanistic pathway of carboxypeptidase (pathway B).

embedded, like a 'pearl in a necklace'. These loops are stabilized by Zn^{2+}, coordinated with two anionic cysteine and two neutral histidine side chains (Figure 8.5). Hence, the geometry around the zinc is tetrahedral in nature. The compact folding of the domain around the metal center, supported by two cysteine and two histidine residues, develops a protruded structure that closely wraps around DNA to recognize a site where information about the synthesis of a specific protein is available (Table 8.4). The three-dimensional structure is comprised of β-sheet folding on the cysteine side and α-helical arrangement from the histidine side. It is worth mentioning that the role of the zinc ion present in RNA synthetase, polymerase, and transcription factors, as a zinc finger, is not catalytic but instead frames the structure of the protein that binds to DNA strands to activate or deactivate genes as per the requirement.

The sequencing order of amino acid present in TFIII A, known as the Cys_2His_2 type, mainly follows a trend: (Phen, Tyr)-X-Cys-$(X)_{2-5}$-Cys$(X)_3$-Phen-$(X)_5$ on one

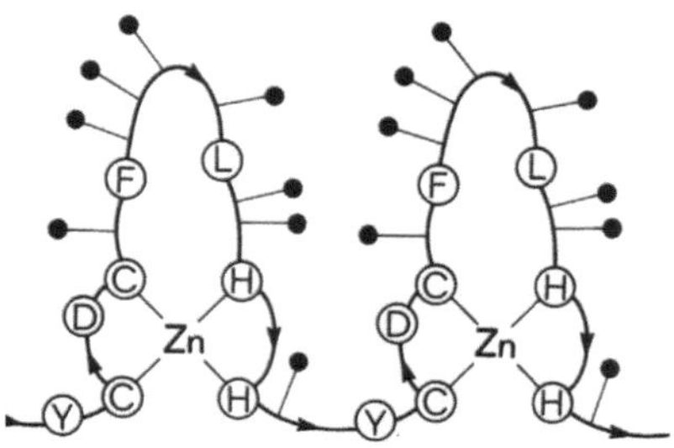

FIGURE 8.5 Diagram of Zinc Finger Transcription Factor IIIA (TFIIIA). It is an archetypal zinc finger protein and a member of a very large multigene family of eukaryotic DNA-binding proteins with multiple functions. Its N-terminal polypeptide carries out sequence-specific DNA and RNA binding, while the C-terminal peptide is involved in the transactivation process. (Reused with permission from Ref. [8]; Copyright (2023) Elsevier.)

TABLE 8.4
Example of Various Transcription Proteins Containing Zinc

Protein	Mol.wt. × 10^{-3}	No. of Zn	Ligands
TFIIIA	40	2	4 Cys & 4 His
TFIIIA	40	9	18 Cys 18 His
GAL-4	17	2	6 Cys
g32P	35	1	3 Cys, 1 His

side and a pattern of -Leu-$(X)_2$-His-$(X)_3$-Leu$(X)_5$ on the other, where x is amino acid. The ending X_5 sequence often has amino acid sequence TGEHP, which forms a flexible linker between multiple zinc fingers in the protein (Figure 8.6). Its coordination determines the folding pattern of protein, and substitution of Cys or His residues with any other ligand results in a loss of function. During the process of binding to DNA, zinc fingers make several contacts. Among these, the most important are hydrogen bonds between the α-helix residues of the Cys_2His_2 protein and the base pairs of one strand of the double-stranded DNA, referred to as the primary strand [8]. These interactions induce conformational changes in the DNA to occur on zinc finger, which include the development of major grooves caused by unwinding of DNA. Hydrophobic and phosphate contacts are other important interactions that take place between zinc fingers and DNA.

Many other zinc-containing structural motifs, such as the GAL4 transcription factor, ADR1 (yeast alcohol dehydrogenase regulatory protein), and ZIF268 (mouse protein)-like proteins, are also well known. Unlike TFIIIA, GAL4 contains binuclear zinc tetrahedrally coordinated by six cysteines, four of which are terminal and the remaining two are of a bridging nature.

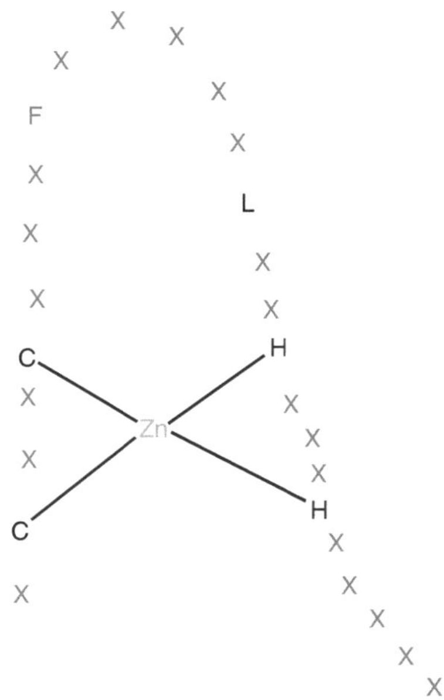

L = Leucine, F = Phenylalaline, Y = Tyrosine, C = Cysteine, H = Histidine

FIGURE 8.6 Schematic representation of zinc finger.

CONCLUSIONS

In summary, zinc fingers change the three-dimensional structure of the DNA strand to which they are attached, opening up through major grooves. Zinc ions might have been chosen due to the following reasons:

1. Their more abundant nature in biological systems.
2. Tetrahedral coordination around the zinc ion shows strong tolerance towards esteemed distortion during contact formation and the grooving process.
3. The presence of a filled electronic configuration (d^{10}) makes the zinc ion inactive toward redox activity, avoiding DNA damage that might be possible with metal centers like iron and copper.

REFERENCES

1. S. J. Lippard, J. M. Berg, *Principles of Bioinorganic Chemistry*, University Science Books, Mill Valley, CA, 1994.
2. R. M. Roat-Malone, *Bioinorganic Chemistry: A Short Course*, 2002, John Wiley & Sons, Inc., New Jersey, Pages 1–365.
3. R. W. Hay, *Bio-Inorganic Chemistry*, Ellis Horwood Limited, Halsted Press, New York, 1984.

4. W. Kaim, B. Schwederski, A. Klein, *Bioinorganic Chemistry: Inorganic Elements in the Chemistry of Life*, 2013, John Wiley & Sons, Inc., New Jersey, Pages 1–426.
5. M. Laitaoja, J. Valjakka, J. Jänis, Zinc coordination spheres in protein structures, *Inorg. Chem.*, 2013, 52, 10983–10991.
6. D. S. Auld, *Handbook of Protelytic Enzymes,* 3rd edition, Academic Press, New York, 2013, 1, Pages 1289–1301.
7. T. Hogg, R. Hilgenfeld, Protein crystallography in drug discovery. In John B Taylor, David J Triggle (Eds), *Comprehensive Medicinal Chemistry*, Elsevier, Amsterdam, the Netherland, 2007, 3, Pages 875–900.
8. D. Neuhaus, Zinc finger structure determination by NMR: Why zinc fingers can be a handful, *Progress in Nuclear Magnetic Resonance Spectroscopy*, 2022, Elsevier, Amsterdam, the Netherland, Pages 130–131, 62–105.

9 Metal Ions in Diagnostics and Therapeutics

INTRODUCTION

Metal ions have played an extensive role in therapeutics and diagnostics owing to their unique chemical properties and interaction with biological systems. Metal-based drugs, commonly referred to as metallodrugs, harness the power of metals in medicinal chemistry and offer chemical functionalities that are not available with pure organic molecules. The use of metals and their compounds for therapeutic and diagnostic purposes constitutes an important class of bioinorganic chemistry called 'Medicinal Inorganic Chemistry'. The utilization of metal ions in therapeutics is not new; they have a long and intriguing history that can be traced back centuries. From ancient civilizations to present times, metals and metal salts have been utilized in a variety of roles in therapeutics. Remedies like the use of copper as a disinfectant to sterilize water, zinc compounds as antiseptics for accelerated wound healing, and gold for treating various chronic diseases as well as promoting longevity have been known since antiquity. There is evidence that ancient Indians, Egyptians, Greeks, Chinese, and Romans had a fair understanding of the medicinal properties of mercury and lead [1]. Cinnabar, a form of mercury sulfide, has been extensively used in traditional Chinese medicines. Mercury compounds were used as anti-syphilis, antipruritic, antiseptic, anti-inflammatory, and diuretic medicines [2].

The twentieth century witnessed one of the first synthetic antimicrobial metallodrugs called Salvarsan. Prior to Salvarsan, most metals or simple metal salts were used in therapeutics. Salvarsan, a mixture of 3-amino-4-hydroxyphenyl-arsenic(III) compounds, also known as arsphenamine or compound 606 [3], was introduced by Paul Ehrlich in 1910 as a remedy for syphilis, a sexually transmitted disease. The number 606 represents the 606th preparation tested in Ehrlich's laboratory. Ehrlich's idea behind Salvarsan was to develop a Magic Bullet, which would kill invading microorganisms without affecting the host cell [4]. Despite its long history and importance, the chemical structure of Salvarsan is still contentious. The initial structure assigned by Ehrlich consisted of an As=As double bond in Salvarsan; however, a few other prominent scientists believe that such an As=As double bond could be stable only in molecules with extensive steric protection. Instead, its structure is proposed to consist of a mixture of tri- and penta-cyclic units of arsenic, which can gradually release $RAs(OH)_2$, possibly imparting the

DOI: 10.1201/9781003432333-9

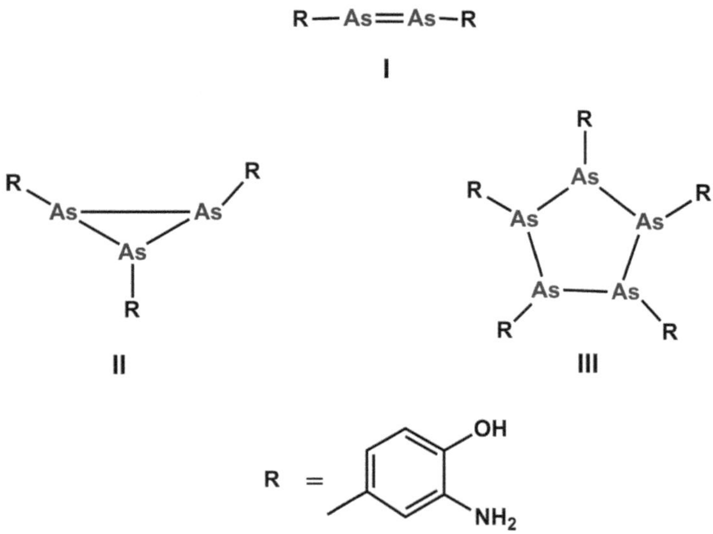

FIGURE 9.1 Structure of Salvarsan (I) proposed by Ehrlich, (II and III) cyclic structure which upon oxidation releases $RAs(OH)_2$. (Redrawn from Ref. [5].)

antisyphilis property to Salvarsan (Figure 9.1) [5]. The discovery of Salvarsan was an important event in syphilis treatment, which was adopted across the globe owing to its effectiveness and relatively low toxicity. Though Salvarsan fell short of being an ideal Magic Bullet due to its side effects, its success paved the way for modern medicine based on synthetic drugs that could be employed to treat disease.

Subsequently, a significant number of metal-containing compounds have been explored for therapeutic and diagnostic purposes. For example, platinum, gold, and ruthenium compounds in cancer therapy [6], gold complexes for arthritis remedies [7], and many other metal-containing compounds for antibacterial, antiviral, anti-inflammatory, and anti-neurodegenerative treatments [8]. Then there are metals such as technetium-99m, gallium-68, and copper-64, whose compounds are administered in radio-diagnosis, as well as gadolinium compounds used as contrast agents in magnetic resonance imaging [4]. A few representative metal-based compounds investigated in therapeutics and diagnostics are given in Table 9.1. From there, it is evident that medicinal inorganic chemistry employs essential as well as non-essential (many of them toxic) elements for medicinal use. Figure 9.2 presents a medical periodic table showing essential elements for humans, medical radioisotopes, and elements used in therapy or diagnosis.

UNIQUE FUNCTIONS OF METAL IONS IN MEDICINE

Versatile electronic, optical, and structural features, including variable oxidation states, tunable coordination geometries, and coordination numbers associated with metal complexes, have hugely benefited metal-based therapeutics.

TABLE 9.1
Some Common Metal-Based Therapeutics and Diagnostics

Metal	Metal-Based Compounds	Therapeutics/Diagnostics
Pt	Dicycloplatin	Liver cancer
	Cisplatin	Lung cancer
	Oxaliplatin	Colorectal cancer
	Dinuclear platinum complex	Various cancers
Au	Auranofin	Rheumatoid arthritis and anticancer
	Gold dithiocarbamate/phosphine	Antitumor activity
	Au nanoparticles	Drug delivery and imaging
Ag	Silver sulfadiazine	Burn treatment
	Silver diamine fluoride	Arrest dental cavities
	Ag nanoparticle	Antimicrobial properties
Cu	Copper histidinate	Menkes disease
	Copper benzimidazole complex	Anticancer
	Cu(II) indomethacin	Anti-inflammatory
Zn	Zinc acetate	Zinc supplement
	Zinc pyrithione	Antifungal and antibacterial
	Zinc lozenges	Remedy for common cold
	Zinc oxide nanoparticles	Antiseptic properties
Ru	NAMI–A, KP10109, RAPTA–C	Anticancer
Fe	Sodium nitroprusside	Vasodilator
	Defroxamine	Antimicrobial activity
	Iron polymaltose	Anemia
	Ferumoxytol and Ferumoxides	Iron-based contrast agents
	Functionalized Fe_3O_4	Drug delivery and hyperthermia
Mo	Tetrathiomolybdate	Breast cancer
Co	Methylcobalamin, cyanocobalamin, and hydroxocobalamin	Supplement for Vitamin B12
	Cobalt(III) bis(2-methylimidazole) Acacen complex	Adenoviral conjuctivitis
V	Bis(glycinato) oxovanadium (IV)	Antidiabetic
Zr	Zirconium glycinato	Antiperspirant
Ti	Titanocene dichloride, bis(β–diketonato) Ti(IV)	Anticancer
As	Salvarsan	Effect against syphilis
	Darinaparsin	T-cell lymphoma
	Trisenox	Acute leukemia
	GSAO	Advanced therapy-resistant solid tumors
Sb	Meglumine antimoniate and sodium stibogluconate	Anti-parasitic disease leishmaniasis
Bi	Bismuth subsalicylate	Gastrointestinal ulcer
Li	Lithium carbonate and lithium citrate	Manic depression and bipolar disorder
La	Fosrenol (lanthanum carbonate)	Hyperphosphatemia

(Continued)

TABLE 9.1 (*Continued*)
Some Common Metal-Based Therapeutics and Diagnostics

Metal	Metal-Based Compounds	Therapeutics/Diagnostics
Gd	Magnevist, MultiHance, Dotarem, Vasovist	Imaging of brain, spine, liver, and blood vessels
Tc-99m	Tc-99m based several nuclear medicines	Nuclear medicine imaging
Ga-68	68Ga-DOTATATE	Positron emission tomography (PET) (neuroendocrine tumor)
Cu-64	Copper(II) (diacetyl-bis(N4-methylthiosemicarbazone)	PET scan for detecting hypoxia region of tumors
Ac	Actinium-225-Labeled Humanized Anti-CD33 Monoclonal Antibody HuM195	Leukemia myelodysplastic syndrome

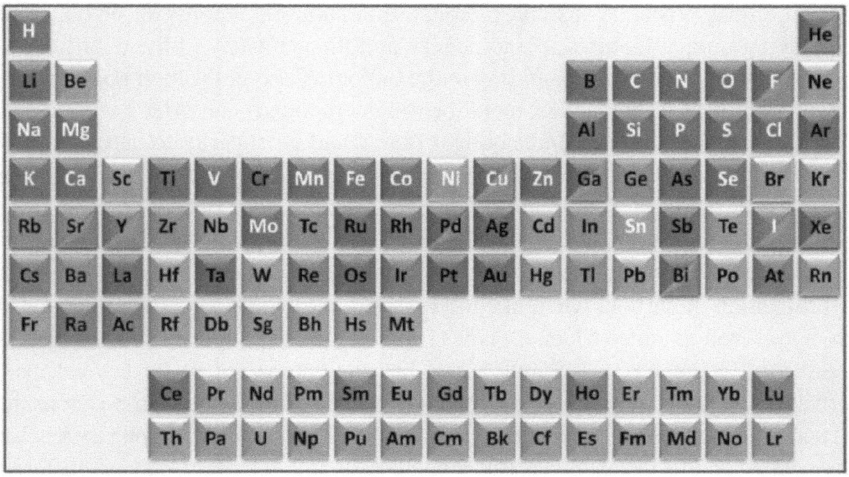

FIGURE 9.2 A medical periodic table revealing essential elements for humans in white font, medical radioisotopes in green fill, elements used in therapy in blue fill, and elements used in diagnostics in orange fill. (Adopted from Ref. [9]. This article is licensed under the Creative Commons Attribution License.)

Furthermore, the intriguing chemistry exhibited by metal-based drugs in terms of ligand substitution reactions, metal- and ligand-based redox behavior, and the catalytic role of metals has made metallodrugs an indispensable tool in modern-day therapeutics. Organic-based therapeutics is not as versatile as metal-based therapeutics. Some metal ions produce biological activity by forming direct covalent bonds with the target site, and in some cases, metal complexes carry the functional ligand to the site of action [10]. In certain cases, the

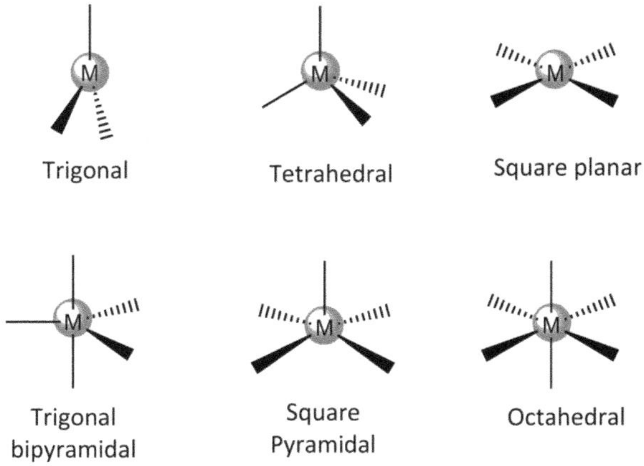

FIGURE 9.3 Possible coordination geometries adopted by a metal ion.

generation of metal-ion-mediated reactive oxygen species can cause apoptosis or cell damage. Metal ions can modulate the catalytic activity of enzymes by employing unique features like ionic size and the geometry of the coordination sphere, which also involves binding to the biological target through non-covalent interactions [11]. Redox-active, metal-bearing compounds can offer a unique platform for drug design that can be quite different and far superior to purely organic drugs. Additionally, metal-ion-bearing scaffolds exhibit a high degree of structural diversity in terms of coordination numbers ranging from two to seven or even higher and geometries, as shown in Figure 9.3. In contrast, the scope of structural variations in organic molecules is rather limited as they primarily rely on the hybridization of carbon, wherein sp-hybridized carbon displays linear geometry, sp^2 hybridized assumes trigonal planar geometry, and sp^3 hybridization results in tetrahedral geometry [12]. Surely, metal-bearing complexes go far beyond these structural variations observed in organic moieties. Further, when we consider structural diversity and stereochemistry together, the structural complexity reaches the next level. For instance, a metal ion in octahedral geometry with six different ligands can exhibit 30 different stereoisomers. These structural variations add new dimensions to modulating the chemical reactivity of metal-containing compounds, which proves to be very crucial for biological phenomena [12]. Considering the impact of ligands on the thermodynamic and kinetic stability of metal complexes, the introduction of a suitable ligand into the coordination sphere can fine-tune the biological action of a metallodrug. Moreover, metal complexes exhibiting luminescence and magnetic behavior could be valuable diagnostic tools. Complexes bearing radioactive metals are already proven assets in diagnostics and therapeutics.

Metallodrugs comprise several components that can serve therapeutic purposes. Hambley et al. have classified metallodrugs into different categories based on the function of the metal and ligand moiety. The first and second categories comprise metal compounds that are active in their inert and reactive forms, respectively. The third category complexes serve as radiation enhancers while the fourth category includes compounds of radioactive metals. The fifth category describes complexes in which the metal or its biotransformation product is active. The last two categories include complexes where only the ligand is active or a certain fragment is active.

According to a recently conducted study on antimicrobial activity, metal-bearing compounds display around 10% higher activity toward ESKAPE pathogens compared to purely organic molecules [13]. Additionally, metallodrugs often function as prodrugs, which provide the opportunity to be selectively activated at the target site by substitution/dissociation of labile ligands, ring opening in response to changes in pH, changes in oxidation state, or by external stimuli such as the application of light, heat, or radiation [10].

Despite offering so many unique advantages and promising results, metal-based scaffolds are less investigated, indicating there is still a lot of space available for exploration. The reduced screening and clinical trials of metallodrugs could possibly be because, firstly, of the inability of metal-bearing compounds to follow the guidelines provided for organic molecules to exhibit drug-like properties. For instance, molecules with MW > 500 are predicted to have poor absorption and permeation, as per Lipinski's Rule of Five [14], which is not applicable to metal-bearing compounds, particularly metals belonging to the third transition series (e.g., gold-containing Auranofin has MW ~678 Da) [10]. Instead, molecular volume could be a more suitable guideline for metallodrugs. Secondly, the use of metallodrugs since antiquity has been largely driven by empiricism rather than rational design which includes the identification of the target site and mode of activation and elucidating the molecular mechanism. Therefore, to exploit the full potential of metallodrugs, it is imperative to recognize their target sites, such as deoxyribonucleic acid or proteins, and also to understand the various strategies available to selectively activate the metallodrugs at the target site.

POSSIBLE BIO-TARGETS FOR METALLODRUGS

Most of the research related to metallodrugs has focused primarily on DNA as a target. However, as our understanding has increased, new targets have also been explored in addition to DNA. The quest for new targets arises from the fact that during therapy, the DNA of normal cells will also be targeted along with that cancerous cells, leading to the killing of normal cells as well. Additionally, there is a significant chance that cells can develop resistance to drugs that have a single biological target. For this reason, the identification of new drug target sites is of paramount importance. In this section, some of the potential targets for therapeutics will be discussed.

Direct Covalent DNA Binders

For cisplatin and other platinum-bearing anticancer drugs like carboplatin, oxaliplatin, lobaplatin, etc., nuclear DNA is the target. They form strong adducts with DNA, which is a crucial step in their anticancer activity [15]. Since cisplatin-based compounds function through the formation of adducts with DNA, they are expected to produce similar kinds of biological consequences. However, the protein recognition of the resulting drug-DNA adduct relies upon the nature and structure of the ligand [16]. It is also observed that linear multinuclear platinum complexes bearing a high positive charge exhibit stronger initial interactions with DNA and are more flexible in comparison to mononuclear platinum complexes [17]. Despite this, they failed in clinical trials due to their instability in blood and their ability to cause severe DNA distortion or significant unwinding of the helix, which is indeed observed with mononuclear platinum complexes.

Non Covalent DNA Intercalators

Unlike covalent DNA binders, which form irreversible bonds with DNA, non-covalent DNA binder metal complexes reversibly associate with the DNA helix. Owing to their appropriate shape and polarity, these metal complexes can either function as intercalators or groove binders [18]. Intercalators/groove binders interact with specific sequences of DNA through hydrogen bonding or hydrophobic interactions, inducing structural and, consequently, functional changes in DNA, thereby interfering with DNA replication, transcription, and other DNA-mediated processes. Apart from intercalation and groove-binding modes, highly charged polynuclear metal complexes like triplatinNC can interact strongly with the phosphate backbone of DNA, forming a phosphate clamp and subsequently impeding DNA-mediated processes even in the absence of direct bond formation with DNA bases [9]. These non-covalent DNA binders play a vital role in modern medicine by targeting DNA and obstructing cellular processes. However, their therapeutic potential and selectivity depend on careful design and modulation.

Simultaneous Binding and Intercalation

These are a unique class of therapeutic agents that possess the ability to bind to DNA while also intercalating between the DNA base pairs. These include metal complexes with extended arene systems, wherein the metal forms a direct bond with DNA while the aromatic ligand is exploited for intercalation. This simultaneous binding mode can also be achieved in metal complexes containing σ-bonded side arms. The platinum-ethidium complex is an example of this category, which exerts its pharmacological effects by simultaneously binding to DNA and intercalating between the base pairs [19]. This simultaneous interaction produces strong structural modifications by increasing the length and unwinding the DNA helix. Other prominent examples from this class include biphenyl ruthenium(II) complexes, cyclopentadienyl iridium(III) complexes, etc.

G-Quadruplex Binders

These are a special class of therapeutic agents that specifically interact with stacked four guanine bases (G-tetrads) known as G-Quadruplexes. Such G-rich strands are often observed in human genes, such as telomerase, which maintains cell division. G-Quadruplex binders have a planar and aromatic core that interacts with the G-rich strand and inhibits the enzyme telomerase, thus acting as an anti-cancer agent. Cu(II) complexes with salphen derivatives or terpyridine-based ligands have displayed the ability to function as effective G-Quadruplex binder.

Targeting Specific DNA Sequence

To overcome side effects and drug resistance, specific recognition of DNA sequences is an elegant strategy [9]. Indeed, an ideal metallodrug molecule should target DNA very selectively. This can be achieved by involving ligands effective in identifying specific DNA sequences. Examples in this category include platinum(II) complexes derived from linear and hairpin polyamide ligands [20]. Pt complexes involving oligonucleotides have also displayed DNA sequence recognition ability while retaining cross-linking features [9]. Another attractive strategy is to incorporate proteins capable of recognizing specific sequences. This strategy is inspired by organic drugs, wherein the organic molecule forms a ternary complex by binding with both proteins and a DNA strand with a particular sequence [9].

Targeting Proteins

For a long time, DNA was the only recognized target for metallodrugs. However, owing to the crucial role rendered by proteins in cellular processes, proteins have also emerged as an attractive target [21]. Targeting proteins not only produces the desired therapeutic effects but also exhibits reduced side effects compared to conventional DNA-targeting therapies. With the emergence of highly sensitive bio-analytical techniques, several potential protein targets, such as kinases, mitochondrial proteins, hormones, etc., have been identified [21]. Several recently conducted studies have shown that ferrocenyl organometallic complexes decorated with anti-androgen hormone compounds are effective for androgen-dependent prostate cancer [9]. Ferrocenyl compounds incorporating tamoxifen, a selective estrogen receptor modulator, are found to be effective against breast cancer cells. The anticancer properties exhibited by ferrocenyl derivatives have been attributed to the redox activity exhibited by the ferrocenyl group and the antiestrogenic properties of the tamoxifen group.

Another potential target for metallodrugs is mitochondrial proteins. Mitochondria, often referred to as the powerhouse of the cell, perform crucial roles in energy production, cellular metabolism, and cell survival. It has now been understood that mitochondria play a vital role in cancer expansion by providing energy to cancerous cells, from redox modulation to regulating cell death pathways [22]. Targeting mitochondrial proteins presents a promising strategy for therapeutic interventions. Potential mitochondrial protein targets include electron transport chain complexes involved in oxidative phosphorylation and ATP synthesis, permeability transition pore regulators that control mitochondrial membrane permeability, mitochondrial DNA maintenance proteins that, upon interaction with metallodrugs, lead to mitochondrial dysfunction, and mitochondrial membrane proteins responsible for membrane potential. Several arsenic-based metallodrugs have been explored to target mitochondrial proteins. Additionally, cationic lipophilic complexes of gold(I) with phosphine or carbenes or ruthenium(II) complexes have been found to disrupt mitochondrial function, impair mitochondrial membrane potential, and thereby induce apoptosis [9]. Targeting mitochondrial proteins is associated with fewer side effects, as metallodrugs can be tuned to selectively accumulate within the mitochondria and interfere with energy production, oxidative stress, cell death pathways, and cellular metabolism.

Likewise, kinase proteins have also emerged as a potential target for metallodrugs. Kinase proteins are responsible for the transfer of phosphate groups to proteins, which activates their function. In this way, kinase proteins regulate cellular functions. Similar to mitochondrial proteins, kinase proteins are closely involved in the development and progression of cancer cells, making them an attractive target for therapeutics. Several rigid, bulky, and inert octahedral complexes bearing metal ions such as Ru(II), Os(II), Rh(III), and Ir(III) can function as kinase inhibitors. These complexes particularly target active sites by adopting rigid shapes [23].

Targeting Metalloenzymes

The crucial role played by metalloenzymes in biological activities has been recognized for a very long time. The ubiquitous role of metalloenzymes makes them a cause of the development and progression of many diseases. This could be due to the overexpression or enhanced activation of metalloenzymes. This makes metalloenzymes an attractive target for therapeutic purposes [24]. the use of metal-binding pharmacophores to inhibit the function of metalloenzymes is an attractive area of research. Several metallodrugs, such as Pt(IV) complexes, have displayed metalloenzyme inhibition properties either by adopting unique 3D structures, incorporating into metal-binding pharmacophores in metalloenzymes, substituting metals from the active site of metalloenzymes or utilizing the intriguing photophysical/redox properties of metal complexes [25].

ACTIVATION OF METALLODRUGS/PRODRUG MOLECULE

The prodrug approach has recently gained considerable attention as an alternative to conventional metallodrug therapy because of its appealing systemically reduced toxicity. However, prodrug molecules often require activation for their action. There are several ways that can be employed to activate metallodrugs. The activation of metallodrugs is a crucial step that determines the interaction of the metallodrug with the biological target and ultimately its therapeutic efficacy. The mode of activation generally depends on the chemical structure and the intended mechanism of action of metallodrugs [10]. This section of the chapter sheds light on the prominent modes of activation employed by metallodrugs.

Activation upon Ligand Substitution

Hydrolysis of metal complexes is one of the most common modes of activation of metallodrugs, particularly transition metal ion-based drugs, which generally do not have high water-ligand exchange rates. This inertness toward ligand substitution with water, in heavier transition metals, has been exploited during drug design. The choice of ligands attached to the metal center in metallodrugs also provides an excellent opportunity to control the reactivity of the metallodrug toward ligand substitution. One of the fine examples of a metallodrug in this category is Cisplatin. Square planar Cisplatin $[PtCl_2(NH_3)_2]$ is activated *in situ* by hydrolysis, leading to the formation of mono- and diaquated species. These aqua-substituted species can bind effectively with DNA. However, the hydrolyzed drug molecule has a strong affinity toward soft cysteine thiolate groups due to the presence of the soft Pt(II) metal ion. Such interactions have been the cause of the adverse side effects of this drug molecule [26]. Carboplatin, a second-generation platinum drug that displayed reduced nephrotoxicity, has addressed this issue by modulating the reactivity of the drug molecule through the choice of ligands. In Carboplatin, labile chloride ions were replaced by a relatively inert chelated dicarboxylate group, which slowed down the rate of hydrolysis and reduced the side effects [26].

Likewise, many other potential drugs in different phases of clinical trials are also activated by the hydrolysis mechanism. These drugs include titanocene dichloride $[TiCp_2Cl_2]$ (Cp is cyclopentadienyl), half-sandwich complexes of Ru(II) and Os(II) derived from η^6-arene ligands, Rh(III) and Ir(III) complexes with cyclopentadienyl ligands, etc. [10]. Anticancer studies of these complexes have shed light on the vital role played by the ligands in the rate of hydrolysis and, in turn, the antiproliferative activity of the complex. For instance, it has been found that substitution of Cl with I or N_3 in $[(\eta^6\text{-biphenyl})Os(en)Cl]$ (where en=ethylenediamine) results in a decrease in the hydrolysis rate. On the other hand, functionalization of the arene ring with electron-donating groups facilitates the hydrolysis process. Interestingly, substitution of the neutral *en* group with an anionic *acac* (acetylacetonate) moiety significantly accelerates the rate of hydrolysis to produce aqua-substituted complexes, which eventually transform into hydroxo-bridged dimer complexes. The latter molecules owing to the presence of inert ligands like hydroxo-bridged species, are ineffective against cancer cells [27]. It has also been inferred that the presence of σ-donor ligands promotes hydrolysis, while π-acceptor ligands slow the rate of hydrolysis and lead to the formation of acidic aqua complexes. Like other 5d series elements, Ir complexes such as $[Ir(H_2O)_6]^{3+}$ are inert. The ligand substitution rate can be increased by the introduction of a cyclopentadienyl group in the complex. Consequently, $[(\eta^5\text{-Cp})Ir(phpy)Cl]$ (where phpy stands for chelated phenylpyridine) undergoes rapid hydrolysis of the chloride ligand. However, substituting the chloride ligand with a neutral monodentate pyridine ligand significantly slows the rate of hydrolysis and displays the ability to produce a significant level of reactive oxygen species in cancer cells, leading to apoptosis [28].

Stimuli Based Activation

This type of activation mechanism is particularly relevant for prodrugs or stimuli-responsive metallodrugs. The prodrug is an almost sixty-year-old concept, which describes prodrugs as inactive molecules that undergo biotransformation in vivo upon a particular trigger and release the active parent drug moiety, which can deliver the desired therapeutic effect. The idea behind the prodrug concept is to enable target-specific drug delivery in order to circumvent the interaction of the drug species with undesired biomolecules, reduce side effects, and improve the physiochemical, physiological, and bioavailability properties of the actual drug molecule [29]. The prodrug concept has become an established tool for both drug discovery and development.

However, to make the prodrug strategy an effective tool, appropriate stimuli are required to activate the prodrug [30]. The stimuli can be categorized as triggers: (i) those within the physiological environment, such as pH changes, redox fluctuations, etc., and (ii) those present outside the physiological environment, such as light, temperature, magnetic fields, ionizing radiation, etc.

Stimuli Which Are within the Physiological Environment

This class of stimuli is also called endo-stimuli. Such stimuli are of great importance for anticancer treatment. Cancerous cells differ significantly from normal cells in terms of their hypoxic or reductive microenvironment, enhanced concentration of glutathione (GSH), acidic extracellular space, and increased levels of reactive oxygen species. These distinct physiological markers associated with cancerous cells can serve as effective stimuli for prodrugs to exert therapeutic effects [29]. Besides these, specific enzymes are overexpressed in states of illness such as inflammation, cancer, etc. The concentration difference of particular enzymes in normal cells and affected cells can also be used as stimuli to activate the prodrug. By exploiting the above-mentioned abnormalities as triggers, a large number of metallodrugs have been developed [29]. Some of the important stimuli are described here.

Redox Stimuli

A large measurable difference in the redox environment has been observed between healthy cells and abnormal cells. Various reducing agents like NADPH, nitroreductase, glutathione, alkaline phosphatase, etc., are present in cancer cells. For instance, glutathione and glutathione disulfide (GSH/GSSG), a major redox couple responsible for regulating the concentration of reactive oxygen species, are approximately four orders higher in tumors than in normal cells. A higher concentration of GSH leads to a highly reductive environment inside tumor cells. Since transition elements such as Pt, Ru, Co, Cu, etc., can exhibit variable oxidation states, compounds of these metal ions can undergo redox reactions. Consequently, the redox environment can be a useful trigger for prodrugs based on redox-active metal centers [31]. An appropriate example of this class is Pt(IV) complexes, which serve as efficient prodrugs. Several Pt(II) drugs such

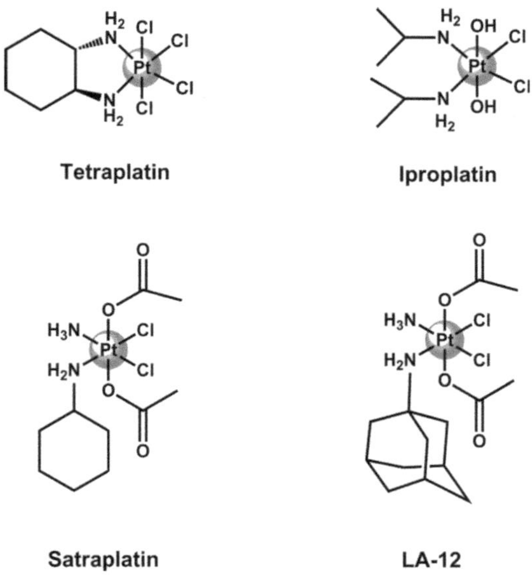

FIGURE 9.4 Pt(IV) prodrugs projected as potential anticancer agents.

as cisplatin, oxaliplatin, etc., are used as anticancer drugs; however, they suffer from severe side effects, drug resistance, and intrinsic toxicity. To address these issues in an effective manner, octahedral Pt(IV) low-spin complexes with inert kinetics were proposed as prodrugs [32]. The inert kinetics of these complexes is particularly useful in minimizing side effects. Studies have revealed that Pt(IV) complexes can be transformed into the active Pt(II) form in the reducing atmosphere available at tumor sites. In contrast, reduction also occurs in normal cells, but the reduced species are rapidly re-oxidized to afford deactivated moieties. The reduction process is accompanied by the elimination of the ligands situated in the axial position. Some of the Pt(IV) compounds, which are derivatives of cisplatin, as shown in Figure 9.4, have shown great promise and have progressed to clinical trial stages [29]. Pt(IV) prodrugs can be functionalized with bioactive ligands at the axial position, which upon reduction can be released in situ to produce the active Pt(II) drug as well as the bioactive ligand, which is expected to enhance the efficacy of the drug. Consequently, several Pt(IV) prodrugs decorated with bioactive molecules have been investigated for their therapeutic effects, including: Estrogen, which makes cancer cells more sensitive to cisplatin. Dichloroacetate, which exhibits anticancer properties by disrupting mitochondrial metabolism β-Cyclodextrin, which bestows the drug with high solubility and stability. Vitamin E analogs, which induce mitochondria-mediated apoptosis. Indole compounds, which enhance cellular oxidative stress. Many more such functionalized Pt(IV) prodrugs have been explored [10].

As already stated, tumor cells have a higher concentration of GSH, leading to a state of hypoxia (a condition where oxygen is insufficient) inside tumor cells. In this state, the levels of reactive oxygen species (ROS), which can cause cell death, are minimal. If the concentration of GSH can be reduced inside tumor cells, it will lead to the generation of ROS. Alternatively, if some reactions can directly produce ROS inside tumor cells, it can result in targeted cell killing. This can be achieved using prodrugs activated by the reductive addition of GSH to form GSSG, which eventually elevates ROS levels. Ru(II) complexes with η^6-arene ligands have demonstrated this mechanism of action [10]. Another possibility arises when metal complexes are directly exploited for the overproduction of ROS in cancer cells. Ferrocene derivatives are one such class of complexes that undergo Fenton-type reactions in cells, leading to the excess production of hydrogen peroxide and subsequently hydroxide radicals. These radicals can directly attack DNA, causing apoptosis [10].

pH-Based Stimuli

Variation in pH across different cellular compartments indicates an acid-base homeostasis in the human body. This difference in pH has been commonly used to activate pH-responsive metallodrugs and delivery stimuli for targeted drug delivery. Further, it has been increasingly recognized that a significant pH difference exists between cells under altered pathological conditions such as cancer, inflammation, and normal cells. Due to this, pH stimuli have emerged as an effective strategy for the activation of prodrug molecules. The design of pH-responsive prodrugs is based on either introducing pH-labile chemical bonds or using pH-sensitive carriers [29].

Thioplatin is one such compound that displays anti-tumor activity without severe side effects. O-dithiocarbonic acids in thioplatin remain attached to the Pt center, forming a chelated ring at higher pH. As a result, the Pt is not available to form bonds with the nitrogen atoms of DNA. It is important to note that sulfur (S), compared to nitrogen (N), has a higher affinity for Pt. However, at slightly lower pH, protonation of the dithiocarbonic acid ligands occurs, followed by ring opening, leading to the formation of an active aqua platinum complex, as shown in Figure 9.5 [33]. This protonated form of the drug exhibits anti-cancer activity.

In addition to the above strategy, pH-responsive drug delivery systems have also emerged as effective tools to enhance therapeutic potency and minimize side effects. The design of a pH-responsive drug delivery system involves metallodrugs encapsulated by pH-sensitive carriers. In this context, pH-responsive polymers capable of undergoing protonation or deprotonation under physiological pH conditions have attracted considerable attention in various cancer therapy fields. One such strategy is to load cisplatin into a polymer containing carboxylate groups. The chloride ions of cisplatin will be replaced by the carboxylate groups of the polymers, only to be eliminated and release active cisplatin under acidic conditions [34].

Enzyme Based Stimuli

Enzymes are vital for the normal functioning of the body. There is growing evidence that in pathologically altered conditions such as inflammation, cancer, and infection, enzymes are present in much higher concentrations compared to normal

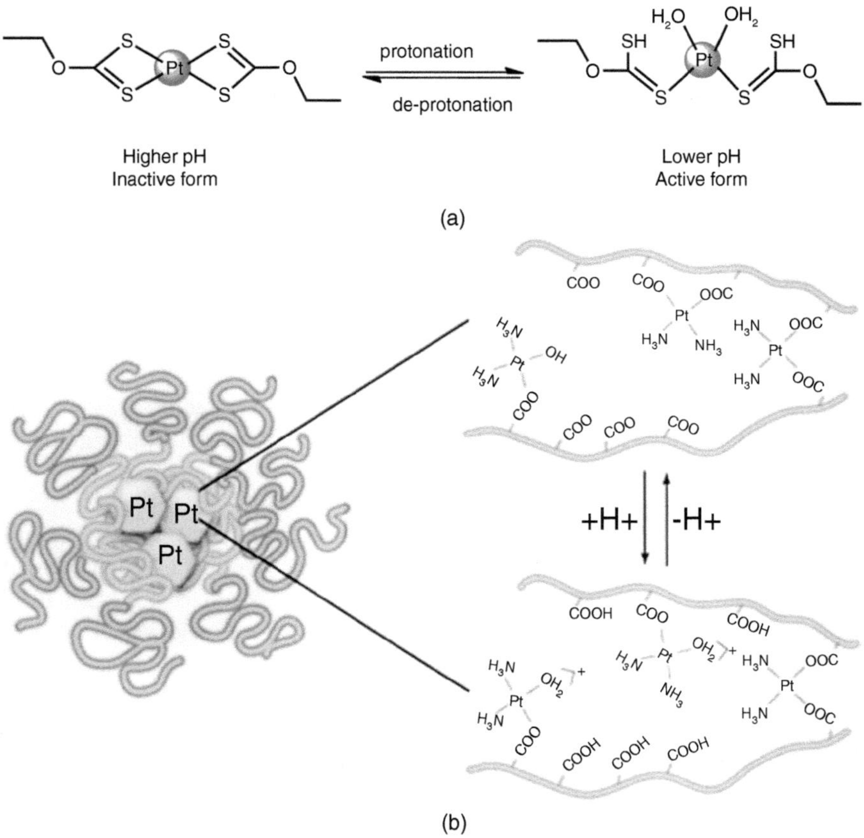

FIGURE 9.5 pH stimuli-based drug activation and drug delivery strategy. (Panel (a) redrawn from Ref. [33] and panel (b) is adapted with Copyright permission from Elsevier, Ref. [34].)

conditions. Due to their overexpression in diseased cells and their characteristic features of specificity and selectivity, enzymes can serve as useful triggers for the activation of prodrug molecules. Various enzymes, such as protease, trypsin, esterase, and phosphatase, have been exploited for drug delivery to abnormal cells. Among these, the cleavage of particular ester or peptide bonds attached to metallodrugs by esterases or proteases is an attractive strategy. For example, a Pt-acridine anticancer agent modified by 2-propanepentanoic ester, which serves as a substrate for human carboxylesterase, has been explored as a prodrug molecule. It has been found that although this prodrug is resistant to hydrolysis, it can be easily cleaved by the human carboxylesterase enzyme, producing the active form of the drug. Furthermore, it has also been observed that the reduced reactivity of the prodrug compared to the parent Pt-acridine complex leads to reduced side effects [35].

Stimuli Which Are Outside the Physiological Environment

Apart from the endogenous stimuli present inside the diseased environment, important exogenous stimuli such as temperature, light, magnetic fields, and ionizing radiation can be used to trigger metallodrugs from outside the body. While the endogenous approach offers no control over the prodrug once it is administered, the exogenous approach provides spatial and temporal control (i.e., the ability to control the timing and location) over the drug molecule [36]. In this section, different exogenous stimuli-based activation mechanisms for metallodrugs will be explored.

Light as Stimuli

Light- or photo-activation offers a powerful strategy for the selective activation of prodrug molecules with high spatial resolution. Photon-mediated activation includes: (i) photodynamic therapy, (ii) photoactivated chemotherapy, and (iii) photothermal therapy.

Photodynamic Therapy (PDT) It is a powerful, clinically approved, minimally invasive therapeutic modality that employs the synergistic action of light of appropriate wavelength, oxygen, and a suitable photosensitizer molecule. PDT indeed offers promising medical potential owing to its intrinsic selectivity and tight spatial and temporal control over the generation of reactive oxygen species. Furthermore, the singlet oxygen, which is the main agent in this technique, has a very short half-life of approximately 40 ns, which limits the therapeutic effect to the site of irradiation, causing minimal damage to surrounding healthy tissue. The efficacy of PDT greatly depends on the photosensitizer. A suitable photosensitizer must possess certain characteristics. First, it should be selectively accumulated in the cancer cells, minimizing side effects. Selectivity can be achieved by incorporating tumor-targeting agents. Secondly, the ideal photosensitizer should be non-toxic in the absence of light, but it should exhibit strong photo-toxicity to produce the desired therapeutic effect. Lastly, and most importantly, the photosensitizer should be excitable by radiation with a wavelength between 600 and 800 nm, which is normally referred to as the therapeutic window. This avoids cell damage from the use of high-energy radiation. Additionally, radiations of higher wavelengths have a larger penetration depth, making them suitable for deep-seated tumors [36].

In PDT, the administration of a photosensitizer is followed by light irradiation, which promotes an electron from the singlet ground state to the singlet excited state of the photosensitizer, from where it can go to the low-lying triplet excited state via intersystem crossing. Once in the triplet excited state, there can be two types of reactions: Type I and Type II. In Type I reaction, an electron is transferred to biological substrates to produce radicals, which subsequently react with molecular oxygen to generate reactive oxygen species. In contrast, in Type II reaction, the photosensitizer in the triplet state can directly react with oxygen to produce singlet oxygen species which trigger oxidative stress, leading to cellular damage and possibly apoptosis [10]. Photosensitizers commonly used in PDT are porphyrins and their derivatives. However, metallodrugs as photosensitizers offer several crucial advantages over organic photosensitizers in PDT. For instance, metallodrugs often exhibit broad and intense absorption spectra, allowing them to absorb visible and near-infrared wavelengths. Metallodrugs are generally associated with longer excited-state lifetimes, which facilitate the triplet state formation via intersystem crossing. Compared to organic photosensitizers, metallodrugs have better solubility in water, resulting in improved cellular uptake. Most importantly, metallodrugs exhibit enhanced stability, are less susceptible to metabolic degradation, and can maintain the integrity of their structure and function for a longer duration compared to organic photosensitizer.

Photoactivated Chemotherapy (PACT) Under the light-activated approach, PACT is an emerging therapeutic modality that blends the advantages of traditional chemotherapy with the spatial and temporal control offered by light stimulation. It involves a metallodrug or prodrug whose interaction with the cell environment is impeded by light-cleavable protecting ligands. Prior to irradiation, both the metal and protecting ligands are engaged with each other, while upon irradiation, cleavage of the metal-ligand bond occurs either by photoreduction, photosubstitution, or a radical mechanism [37]. Light-induced metal-ligand bond cleavage generates bioactive scaffolds that can produce the desired therapeutic effect, as represented in Figure 9.6. PACT functions either by the photo-reduction of the metal complex to generate cytotoxic metal species such as Pt(II) species, and an organic ligand, or metal complexes are activated by photosubstitution of a ligand with a water molecule via a triplet metal-centered excited state, which has a strong dissociative character. Additionally, photo-induced structural transformation of the prodrug molecule to generate the bioactive form is also an effective strategy to produce a therapeutic effect [38]. Being an oxygen-independent mechanism, PACT offers a unique advantage over PDT in terms of its efficacy in the state of hypoxia, usually prevalent in cancer cell.

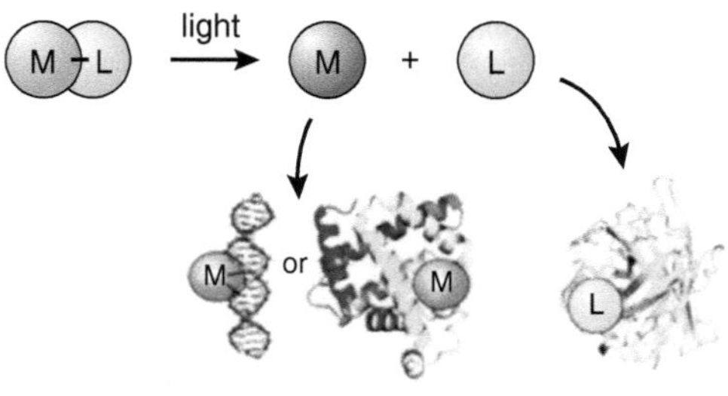

FIGURE 9.6 Light activation of an inert prodrug leading to therapeutic effect. (Copyright permission from Royal Society of Chemistry, Ref. [37].)

Photothermal Therapy (PTT) It is an appealing therapeutic modality that harnesses light for controlled heat-mediated therapeutics. PTT generally employs a photothermal transduction agent (PTA) that facilitates non-radiative conversion of light into heat energy. For an ideal PTA, the material should selectively accumulate in the tumor rather than in normal cells and it should have high photothermal conversion efficiency. Furthermore, PTAs should exhibit high spatial control, i.e., they should confine heat generation to the specific region of interest and thus limit damage to surrounding cells. One of the key considerations in PTA design is the selection of a suitable absorption wavelength. PTAs are usually engineered to absorb light within the tissue-transparent window, which varies from 750 to 1,350 nm. This region is associated with low tissue scattering and improved penetration depth, thus allowing better light absorption and deep tissue heating. Various materials, including both inorganic and organic materials of diverse length scales, have been explored as PTAs for PTT [39]. However, nano PTAs have garnered significant appeal due to their ability to accumulate in tumors through the mechanism of enhanced permeability and retention effect. Nano PTAs have an added advantage over organic molecular PTAs in terms of higher photothermal conversion efficiency. Among nano PTAs, nanocrystals of noble metals such as Au, Pt, and Ag have been extensively explored for their potential as suitable PTAs due to their exceptional resistance to oxidation. In particular, Au nanoparticles have received considerable attention. The absorption properties of Au nanoparticles are strongly influenced by localized surface plasmon resonance (LSPR), which can be very precisely tuned by modifying the size and morphology of Au nanoparticles. (To give a brief idea of LSPR: When metallic nanoparticles are exposed to light, the surface electrons start oscillating collectively, resulting in the formation of resonant electron density waves known as plasmons. The plasmons can be excited at specific wavelengths, displaying strong absorption of light followed by the release of energy in the form of heat through non-radiative decay). Consequently, Au nanoparticles of various morphologies, from nanorods to nanosheets to nanocages, have been explored for PTT [39].

Temperature as Stimuli

Temperature can serve as an effective external stimulus for the activation and release of metallodrugs. One strategy to exploit temperature as a trigger is through the use of thermosensitive polymers. These polymers can exhibit structural transformation in response to changes in temperature, shifting from a shrunken state to a swollen state or vice versa. This structural transition is particularly useful for therapeutic purposes, as it constitutes temperature-responsive systems that take advantage of the temperature differences between diseased and healthy tissues, allowing for the controlled release of metallodrugs at the target location [40]. This is achieved by regulating the temperature around the lower critical solution temperature (LCST) at the diseased site. LCST is the temperature at which the polymer undergoes a phase transition, resulting in a change in its swelling behavior. At temperatures below the LCST, the polymer is in a swollen state and can allow the loading of metallodrugs.

Once the temperature exceeds the LCST, the polymer undergoes structural changes from the swollen to the shrunken state, thereby causing the release of the metallodrug [29]. Since a temperature difference exists between diseased and normal cells, temperature-responsive polymers enable the release of metallodrugs with tight spatial control. Furthermore, the temperature-responsive behavior of the polymers can be modulated by adjusting their composition, molecular weight or chemical structure. This strategy has been exploited for platinum(II) metallodrugs conjugated to thermosensitive polymers derived through cyclotriphosphazenes decorated with alkoxy PEG, which showed LCST behavior [41].

To produce temperature changes, magnetic nanoparticles have emerged as promising heating probes for hyperthermia applications. Among these, carbon-encapsulated magnetic (mostly iron) nanoparticles with core-shell architecture are particularly appealing due to their enhanced stability against oxidation and ease of functionalization with carboxylic groups, which are particularly useful for further complexing with platinum drugs. The platinum drug can be conveniently released in a controlled fashion in response to localized temperature changes produced by magnetic nanoparticles. This strategy opens up new avenues for the development of temperature-responsive systems that can be utilized for bimodal treatment, i.e., hyperthermia and chemotherapy simultaneously [29].

Ionizing Radiation as Stimuli

Radiations such as X-rays and gamma rays can serve as a means to activate metallodrugs. The synergistic action of radiosensitizer compounds with ionizing radiation can produce a therapeutic response that surpasses the effect of each component individually. The radiosensitization of metallodrugs is usually achieved either by enhancing the damage caused by ionizing radiation or by inhibiting the repair and resistance mechanisms [42]. This strategy has been particularly employed in the case of platinum drugs, which are often combined with external beam therapy. It has also been observed that heavy metal-bearing metallodrugs exhibit enhanced therapeutic effects in radiosensitization. Notably, a cyclometallated Ir(III) complex demonstrated the ability to accumulate in mitochondria and exhibit significant radiosensitization in cancer cells. The increased production of reactive oxygen species upon X-ray irradiation is considered to be the reason behind the heightened radiosensitization [10].

METALS IN MEDICINE

Until now, potential targets for metallodrugs have been identified. Although DNA remains a primary and favored target, other biomolecules, such as G-quadruplex sequences and specific proteins (including mitochondrial proteins, kinases, and metalloenzymes), have emerged as potential targets. The activation modes and the influence of endogenous and exogenous stimuli on targeted therapeutic effects have been assessed. This information is vital for the drug design and development of metallodrugs. Metallodrugs have primarily been explored as anticancer agents. However, their applications extend beyond cancer treatment. In the remaining part of the chapter, the diverse applications of metal complexes, including their use as anticancer agents, antiviral agents, antibacterial agents, and their potential in addressing neurodegenerative diseases, will be explored. Furthermore, the role of metal complexes in diagnostics will also be discussed. Additionally, the rising role of nanomaterials, which provide an excellent platform for therapeutics and diagnostics, will also be covered.

Anticancer Metallodrugs

Although the advent of Ehrlich's Salvarsan in 1910 marked the beginning of modern chemotherapy, the discovery of cisplatin in 1965 by Barnett Rosenberg and Loretta VanCamp at Michigan State University revolutionized cancer treatment and ignited extensive research into metallodrugs for anticancer purposes. However, the discovery of cisplatin as an anticancer drug was serendipitous and arose from an entirely different research context. Rosenberg and VanCamp were investigating the effect of electric current on Escherichia coli bacterial cells when they noticed that the application of electric current through platinum electrodes hindered cell division in E. coli. Intriguingly, they discovered that the hindered cell division was not linked to the electric current itself but rather to a compound produced from platinum electrodes.

This compound was later identified as cis-diamminedichloroplatinum(II), commonly known as cisplatin. Subsequent investigations revealed that cisplatin exhibited potent anticancer activity, heralding a new era in cancer chemotherapy. Cisplatin, which revolutionized research on metallodrugs, was synthesized in 1844 by M. Peyrone, and its molecular structure was first explained by the father of coordination chemistry, Alfred Werner, in 1893. The discovery of cisplatin as an anticancer agent has prompted further exploration of platinum-bearing complexes for anticancer properties. Molecular structures of prominent platinum anticancer compounds are presented in Figure 9.7.

At this point, it is also useful to give a brief overview of the devastating disease called cancer. Unlike normal cells, which grow, divide, and die in a regulated manner to maintain healthy tissues, cancer cells display abnormal and uncontrolled cell division, invasion into neighboring tissues, and disruption of their functions. Furthermore, the most concerning aspect of cancer is its ability to metastasize. Cancer cells can break away from the primary tumor and spread to other parts of the body through the bloodstream or lymphatic system. These migrating cancer cells can form new tumors in distant organs or tissues, leading to the spread of the disease.

Mechanism of Action of Cisplatin

Cisplatin and related platinum compounds have been widely used in cancer treatment. These compounds operate via a similar mechanism of action and display their ability to inhibit cell growth and induce cell death. They exert their anticancer effects by binding to DNA and disrupting its structure and function. However, the process is complex and involves multiple steps, making a clear understanding of the mechanism crucial. The first step involves the cellular uptake of cisplatin and its transport near the nucleus, where the target DNA is located. As a neutral

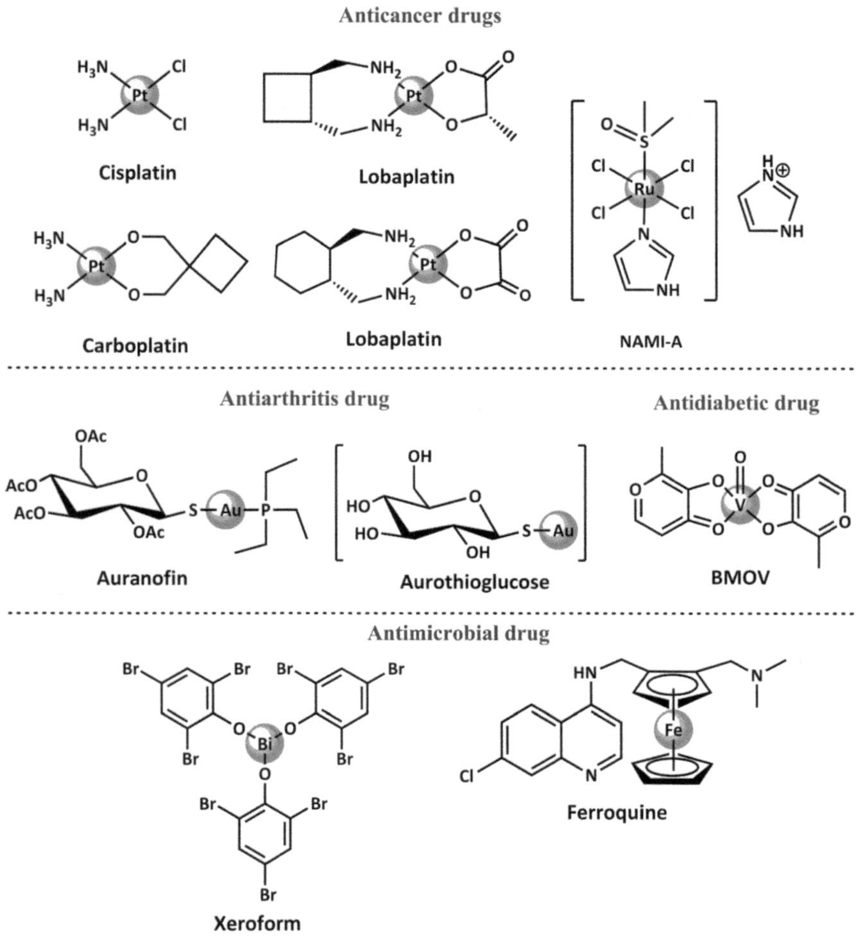

FIGURE 9.7 Molecular structures of common metallodrugs.

molecule, cisplatin can cross the cell membrane by passive diffusion. However, studies have also indicated the involvement of active transport mechanisms in the transport of cisplatin. Notably, the uptake of cisplatin by copper transport proteins has been suggested. Once inside the cell, cisplatin undergoes activation through a hydrolysis process, where chloride ligands are substituted by water molecules. This substitution occurs in a stepwise manner, driven by the low intracellular chloride concentration compared to the bloodstream. The facile ligand substitution is attributed to the square planar geometry of cisplatin. This substitution is a crucial step that generates highly reactive mono-cationic cis-$[Pt(NH_3)_2Cl(OH_2)]^+$ and di-cationic species $[Pt(NH_3)_2(OH_2)_2]^{2+}$. The weakly bound water molecules within the coordination sphere can be replaced by nitrogen atoms on DNA, particularly guanine. After coordinating with one strand, the remaining water or

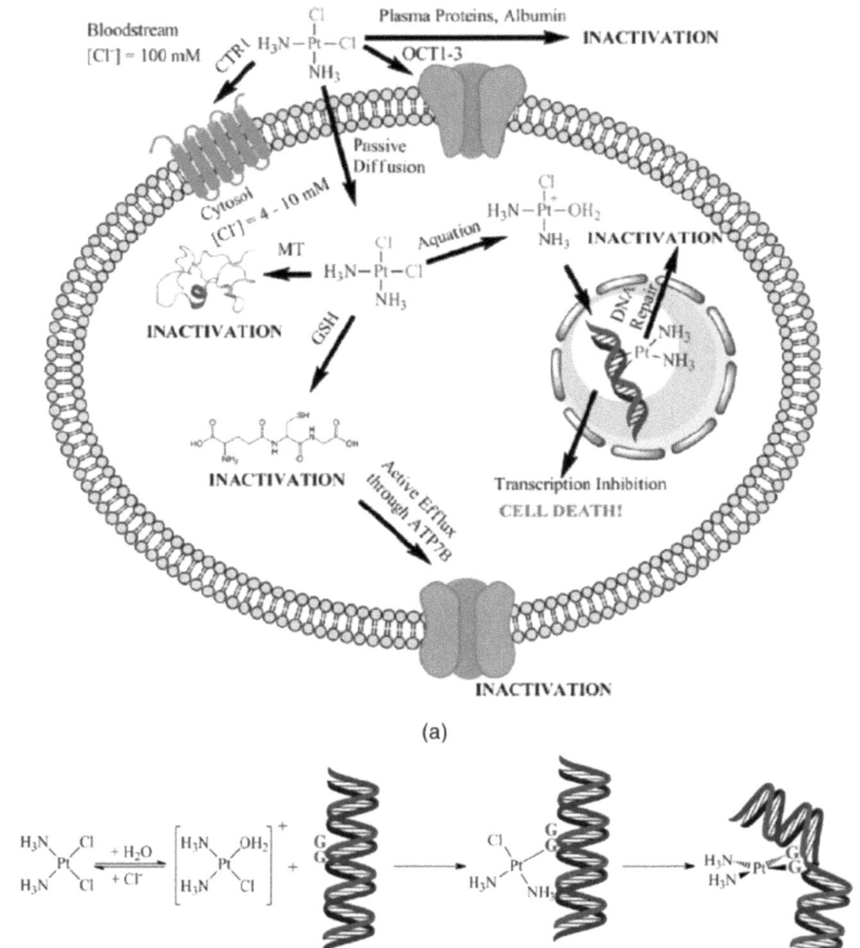

FIGURE 9.8 Schematic representation of mechanism of action of cisplatin; (a) various pathways available for cisplatin before and after penetrating the cell membrane and (b) cisplatin binding to guanine nucleobases of DNA forming a cross-link. (Copyright permission from American Chemical Society, Ref. [43].)

chloride ligand is replaced by the guanine nitrogen from an adjacent strand. This creates a unique-cross linking of two DNA strands by the platinum fragment within the double helix. Such cross-linking interferes with DNA replication and transcription, resulting in DNA damage and eventually cell death. It is important to note that, besides DNA, cisplatin has a high tendency to interact with species bearing soft donor atoms, such as sulfur-containing glutathione and thiocysteine. These intersections can sequester cisplatin and prevent it from reaching the DNA in the nucleus, thereby limiting its effectiveness. The overall mechanism is schematically presented in Figure 9.8 [43].

Interestingly, the therapeutic activity of diamminodichloroplatinum(II) is specific to the cis-isomer (cisplatin). The geometrical arrangement of chloride ligands in cisplatin facilitates its strong binding to DNA. Conversely, the transplatin isomer does not exhibit similar therapeutic activity, possibly due to its deactivation before reaching DNA.

Though cisplatin is highly effective against cancer, it can also cause mutagenesis, where a normal base is replaced by a different base. This could have severe deleterious effects on the health of patients administered with cisplatin. Among the various biological consequences of cisplatin-DNA interaction, one significant consequence is the development of resistance to platinum drugs. The acquired resistance is attributed to DNA repair mechanisms, where the sugar-phosphate backbone on the platinated strand is hydrolyzed, causing the expulsion of the platinated oligonucleotide and facilitating DNA repair. Additionally, glutathione present in the cell binds to cisplatin, reducing its availability to bind with DNA [44]. However, if the cell cannot repair cisplatin-mediated DNA damage, the concentration of pro-apoptotic proteins increases, triggering the release of cytochrome C and leading to apoptosis. The side effects associated with cisplatin have sparked a quest for alternative platinum-based anticancer drugs with a wider range of activity against various tumors, reduced side effects, and efficacy against cisplatin-resistant tumors. Consequently, drugs like carboplatin, oxaliplatin, nedaplatin, and lobaplatin have emerged as potential alternatives. These drugs exhibit fewer side effects compared to cisplatin. Additionally, various platinum(IV) compounds have been explored as promising anticancer agents. Platinum(IV) compounds exhibit distinct properties compared to platinum (II) complexes. Unlike platinum(II) complexes, which generally adopt square planar geometry, platinum(IV) complexes exhibit octahedral geometries. The presence of a saturated coordination sphere in a low spin d^6 configuration imparts enhanced kinetic inertness to platinum(IV) complexes compared to platinum(II) complexes. This intrinsic inertness is particularly beneficial in minimizing undesirable interactions with unintended biomolecules, thereby reducing side effects. As previously mentioned, placing cancer-targeting moieties at the axial positions platinum(IV) complexes presents a n excellent opportunity for targeted action. The unique fundamental properties and intriguing structural features of platinum(IV) complexes, including kinetic inertness, chemical versatility, and opportunities for alteration to achieve targeted effects, make them an appealing area for further exploration [43].

The remarkable success of platinum-based anticancer drugs has been severely challenged by associated toxicity and acquired drug resistance by cells. These limitations have led to the exploration of non-platinum metal complexes, such as those of titanium, iron, copper, gallium, ruthenium, and gold. These metal complexes exhibit different mechanisms of action and toxicity profiles, making them promising alternatives to platinum-based drugs [45].

Ruthenium, Gold and Gallium Based Anticancer Drug

The Three ruthenium (Ru) complexes, namely trans-[$RuCl_4$(DMSO)(Imidazole)](ImidazoleH) commonly called NAMI-A; trans-[$RuCl_4$(Indazole)$_2$](IndazoleH), commonly called KP1019; and KP1339 (a modified form of KP1019 to improve solubility), have exhibited promising anticancer applications and are currently undergoing clinical trials. The molecular structure of NAMI-A is shown in Figure 9.7. The success of Ru compounds in cancer therapy can be attributed to several factors. Firstly, Ru's ability to exhibit both +2 and +3 oxidation states (OS) with equal ease is significant. Unlike Pt, which exhibits different coordination behaviors in its two preferred OS (square planar in +2 OS and octahedral in +4 OS), Ru maintains octahedral geometry in both +2 and +3 OS. This similarity in geometry facilitates the electron transfer process from +3 to +2 OS, which can be challenging for Pt due to required changes in coordination number and interatomic distances. This structural difference between Ru and Pt is believed to contribute to their differing mechanisms of action.

Besides this, another important distinction between Ru and Pt drugs arises from their mode of transportation into cells. While Pt drugs, particularly cisplatin, utilize copper transport proteins for cellular entry, Ru drugs, particularly KP1339 and KP1019, rely on the transferrin pathway. Cancer cells often exhibit increased expression of transferrin receptors, leading to the preferential accumulation of Ru drugs in cancer cells. This targeted accumulation allows for higher drug concentrations and therapeutic effects while minimizing side effects in normal cells. These features distinguish Ru drugs from Pt drugs and provide a possibility for treating tumors resistant to Pt-based therapies. However, it is worth noting that, although the structural characteristics and cellular entry pathways differ between these two drug classes, their primary target biomolecules remain the same, i.e., DNA. They form adducts with DNA, disrupting replication and transcription processes and exerting their therapeutic effects [46].

In contrast to drugs specifically designed to target DNA, other drugs have been developed to target cellular signaling pathways that are overexpressed in cancer cells. Thioredoxin reductase and glutathione reductase are such targets, and phosphole-containing Au(I) complexes have been identified as potent inhibitors of these enzymes, even at nanomolar concentrations [47]. By disrupting thioredoxin reductase function and glutathione reductase, these drugs interfere with redox signaling pathways and induce cytotoxic effects in cancer cells. Gallium is another interesting element that has shown potential in cancer treatment. Being very similar to Fe(III), Ga(III) can bind to transferrin in the bloodstream and subsequently gain entry into cells through transferrin receptor-mediated endocytosis. Ga(III)-mediated cell death involves various processes. Ga(III) disrupts cellular iron uptake and iron homeostasis, inhibiting Fe-dependent ribonucleotide reductase, which is vital for DNA synthesis. Furthermore, Ga has been found to activate the Bax protein, which triggers apoptosis through the release of cytochrome C in mitochondria and subsequent activation of caspase-3, an important enzyme in the apoptotic pathway. Unlike Fe(III), Ga(III) cannot be reduced under physiological conditions. Currently, two Ga(III) compounds are undergoing clinical trials for their anticancer activity [48].

Antiarthritic Metallodrug

Rheumatoid arthritis (RA) is an inflammatory autoimmune disease, the exact cause of which is still not fully understood. The disease begins with the inflammation of the joints, which subsequently leads to progressive joint destruction, causing restricted movement and persistent pain for the patient. The therapeutic value of gold compounds has been known for a very long time, particularly in a practice known as chrysotherapy. In the last century, another application of gold complexes as antiarthritic agents has emerged, and since then, several gold(I) compounds have been administered to halt the progression of rheumatoid arthritis. These include charged, water-soluble complexes such as sodium aurothiomalate, aurothioglucose, sodium aurothiopropanol sulfonate, and sodium aurothiosulfate, which are administered through painful intramuscular injections. However, triethylphosphinegold(I) tetra-O-acetylthioglucose, commonly known as auranofin, is an FDA-approved drug for RA that has exhibited relatively better effectiveness against RA and can be administered orally. The molecular structures of auranofin and aurothioglucose are shown in Figure 9.7. The mechanism of action of these gold compounds as antiarthritic agents is largely unknown. However, the binding of Au(I) to protein thiol groups, which inhibits the formation of disulfide bonds, is thought to be a possible mode of action. Such binding induces the denaturation of proteins and the subsequent formation of macroglobulins [44]. While gold(I) compounds exhibit therapeutic effects against RA, concerns remain regarding their toxicity, slower rate of clearance, and the unclear structure of intramuscular gold solutions. Besides this, the lack of a clear understanding of the mode of action has limited their widespread use. It will be interesting to see whether nanoscience can revive the popularity of gold pharmacology [4].

Antidiabetes Metallodrugs

Diabetes is a metabolic disorder characterized by high blood sugar levels. It occurs when the body is unable to produce an adequate amount of insulin or cannot effectively use the insulin it produces. There are mainly two types of diabetes: Type 1 and Type 2. Type 1 diabetes is an autoimmune disorder that prevents the pancreas from producing insulin, whereas in Type 2 diabetes, the pancreas produces less insulin than required, and the body also becomes resistant to insulin.

Various metal compounds, especially vanadium compounds such as sodium vanadate and sodium orthovanadate, have exhibited promising insulin-enhancing and antidiabetic effects. These compounds have been found useful in managing blood sugar levels in Type 2 diabetes patients. This has fueled extensive research on vanadium-bearing compounds as potential antidiabetic agents. Consequently, bis(maltolato)oxovanadium(IV) (BMOV) and its congener bis(ethylmaltolato) oxovanadium(IV) (BEOV) have emerged as lead compounds, exhibiting increased bioavailability [9]. The molecular structure of BMOV is shown in Figure 9.7. These compounds are believed to activate insulin receptors by inhibiting insulin receptor tyrosine kinase (IRTK)-associated phosphatases, which leads to the enhancement of insulin function. However, they are associated with side effects that affect the kidneys of patients. Furthermore, the wide acceptability of first-generation vanadium complexes, such as BMOV and BEOV, has been challenged by the high doses required to achieve the desired therapeutic effect. To address this issue, researchers have developed second-generation ligand systems, such as bis((5-hydroxy-4-oxo-4H-pyran-2-yl)methylbenzoatato)oxovanadium(IV) (BBOV), which exhibit half the acute oral toxicity compared to BMOV. Further research and development in this area aim to improve the therapeutic efficacy of vanadium compounds while minimizing their side effects [9].

Antimicrobial Metallodrugs

Arsenic-based compounds have been employed as antimicrobial and antiparasitic agents in therapeutics. The development of arsenic-based drugs can be traced back to the use of arsanilic acid commonly called atoxyl, in 1907. The success of arsanilic acid against trypanosomiasis, which causes sleeping sickness, propelled further research by Ehrlich and coworkers on arsenic antimicrobials. Theer focused efforts ultimately led to the discovery of Salvarsan, marking the beginning of chemotherapy. Though most arsenic-based antimicrobials have been replaced by less toxic drugs, Melarsopol [2-(4-amino)-(4,6-diamino-1,3,5-triazin-2-yl)-phenyl-1,2,3-dithiarsolan-4-methanol] remains in use despite its adverse side effects [4]. The exact mechanism of action of Melarsopol is not fully understood; however, it is believed to exert its therapeutic effects in multiple ways, such as the generation of ROS causing oxidative damage, inhibition of key enzymes of trypanosomes thereby disrupting critical metabolic pathways and cellular functions, and binding to DNA, which impedes DNA replication and transcription.

In addition to arsenic, compounds of antimony and bismuth have also been explored as antimicrobial agents. Potassium antimony tartrate has historically been used to combat mucocutaneous leishmaniasis. However, it has been observed that antimony (V) compounds, such as Stibosan, Neostibosan, and Ureastibamine, exhibit lower toxicity compared to antimony (III) compounds [4]. On the other hand, bismuth drugs were primarily used to treat gastrointestinal disorders but are also employed in combination with other drugs to treat Helicobacter pylori (H. pylori) infection. Bismuth formulations, in the form of colloidal bismuth subcitrate or ranitidine bismuth citrate, are used to treat peptic ulcers caused by H. pylori infection. In the treatment of H. pylori, bismuth drugs are often combined with clarithromycin, which is used as an antibacterial drug. Another compound of bismuth, tribromophenatebismuth(III), also called xeroform (molecular structure shown in Figure 9.7), has demonstrated antimicrobial properties and is often used as a substitute for iodoform in wound treatment [49]. Currently, bismuth thiol compounds such as BisEDT have attracted significant attention as antimicrobial agents, particularly in the case of diabetic foot ulcers. These bismuth-based treatments offer potential benefits in managing chronic wounds and preventing infections.

Beyond Group 15 compounds, silver compounds have also been extensively used in wound treatment and infection management [50]. Ointments containing silver sulfadiazine and cerium nitrate-silver sulfadiazine have been utilized to prevent and treat infections associated with second- and third-degree burns [4]. Furthermore, silver ion-embedded textiles are commonly used in medical products such as surgical wound dressing clothes for infection prevention and treatment purposes. In specific cases, silver alginate is being accepted as a drug for the prevention of central line infections in very low birth weight infants. Additionally, iron-containing compounds, such as ferroquine (molecular structure shown in Figure 9.7), prepared from the combination of ferrocene and the anti-malarial drug chloroquine, have been investigated for their anti-malarial properties in cases of resistance against chloroquine.

Antiviral Metallodrugs

While there has not been measurable success in developing antiviral metal compounds, one compound, Bis(2-methylimidazole)[bis(acetylacetone)(ethylenediamine)] cobalt(III), commonly known as Doxovir (molecular structure shown in Figure 9.7), has shown remarkable success in clinical trials. Doxovir has reached phase II trials for the treatment of herpes simplex labialis and phase I clinical trials as a therapy for two viral eye infections. Studies have indicated that Doxovir exerts its antiviral activity by impeding the entry of the virus into the cell by obstructing membrane fusion events. It is suggested that Doxovir may covalently attach to the histidine moiety of zinc finger domains, thereby not allowing the protein to interact with its recognition site, which is crucial for the membrane fusion process. Additionally, it also inhibits cell-to-cell spread [51].

Transition metal oxyanions, also known as polyoxometalate (POM) frameworks, are spherical polyanionic structures containing bridging oxygen atoms. They have shown great potential as antiviral agents, particularly those derived from vanadium, molybdenum, niobium, and tungsten [10]. These compounds are found to inhibit different families of enzymes that are crucial for viral replication. POMs have displayed efficacy against a range of viruses, including the human immunodeficiency virus (HIV), severe acute respiratory syndrome (SARS), influenza A and influenza B, herpes virus, and hepatitis B virus. Imaging techniques have revealed that POMs locate themselves on the cell surface and impede the membrane fusion process, thereby preventing viral invasion [52].

Another metal complex, viz. Ranitidine bismuth(III) compound, is found to be effective against the SARS virus. The Ranitidine bismuth(III) compound inhibits the ATPase activity of the viral helicase protein by binding to the cysteine-rich region of the N-terminal zinc-binding domain of the viral helicase and thus interferes with the SARS replicative cycle [53]. Likewise, auranofin has been found to impede the SARS-CoV-2-virus replication cycle in human cells. Auranofin-treated cells exhibit a significant reduction in SARS-CoV-2-induced cytokines, suggesting that auranofin could be a useful drug to limit SARS-CoV-2 infection. The therapeutic effects of auranofin against viral infection and associated complications are attributed to its antiviral, anti-inflammatory, and antioxidative properties [54].

Zinc is a special metal that has shown huge potential as an antiviral agent. The Zn metallopeptidase angiotensin-converting enzyme 2(ACE2) complex functions as an entry point for SARS viruses. A combination of Zn(II) ions and ionophores has exhibited antiviral activity by obstructing RNA synthesizing activity, thereby impeding the replication and transcription processes of SARS [55]. Ionophores such as pyrithione facilitate the uptake of zinc, thereby increasing the therapeutic effect. Additionally, copper alloys, such as brass and copper-nickel surfaces, are also found to be effective in inactivating the virus. The production of superoxide and hydroxyl radicals on the surface of copper alloys is considered the reason behind the inactivation [10].

Besides their antiviral activity, metal compounds are also used as adjuvants in vaccines to enhance the immune response. Aluminum compounds, such as aluminum hydroxides [Al(O)OH], aluminum phosphates [$Al(OH)_m(PO_4)_n$], and potash alum [$KAl(SO_4)_2{\cdot}12H_2O$], have been considered The 'gold standard' of all adjuvants. The mechanism of action of adjuvants is possibly due to the adsorption of antigens on the surface of the colloidal adjuvant particles, induction of localized necrosis, and dendritic cell activation [56]. Furthermore, mercury compounds, such as sodium-2-ethylmercurithio-benzoate, commonly known as thiomersal, are used as preservatives. Binding of the thiol group in protein structures by the ethylmercurithio cation of thiomersal is considered a key step in their role of impeding enzymatic activity [4].

Metallodrug for Psychiatric Disorder

Bipolar disorder is a kind of psychiatric disorder characterized by mania and elevated or depressed mood, adversely impacting the patient's quality of life and increasing the tendency for suicide. The efficacy of lithium salts, particularly lithium carbonate and citrate, in treating bipolar disorder has been known for almost 200 years. However, the exact mechanism of action of lithium salts has not been elucidated, but it is known to act on multiple levels, which may contribute to its therapeutic functions, including antioxidant properties, regulation of neurotransmission, modulation of cellular and intracellular changes, immunomodulation, and anti-apoptosis characteristics [4]. Additionally, inhibition of glycogen synthase kinase-3 by lithium, which modulates the cellular signaling pathway, may contribute to the neuroprotective properties of lithium in bipolar disorder. Despite the positive effects of lithium, particularly in cases of bipolar disorder, high doses of lithium salts are known to cause side effects, and there exists a very narrow therapeutic window for lithium salts, i.e., a small range between the beneficial therapeutic effect and toxic effect [57].

Metallodrug for Cardiovascular Disorder

Metallodrugs exert their therapeutic effect in cases of cardiovascular disorders by regulating the amount of nitric oxide (NO), which controls vasodilation. In cases of high blood pressure, NO donor metallodrugs cause dilation of blood vessels, thereby promoting increased blood flow throughout the body. Sodium nitroprusside, $Na_2[Fe(CN)_5(NO)].2H_2O$, is a widely used drug that quickly decreases arterial pressure and peripheral resistance, leading to vasodilation. However, one major concern related to sodium nitroprusside is that, along with NO, it releases toxic cyanide, which has a detrimental effect. Therefore, in the quest for an alternative metallodrug to sodium nitroprusside, next-generation NO coordination complexes of Ru and photoactive Fe are currently under investigation [4]. It is expected that these new-generation metallodrugs may exhibit improved therapeutic efficacy and alleviate the side effects associated with sodium nitroprusside. Furthermore, ruthenium NO donor complexes under investigation to address cardiovascular disorder have also shown promise in the treatment of parasitic diseases.

Apart from NO donor complexes, NO absorbing complexes are also beneficial, particularly in cases of toxic shock syndrome where blood pressure is extremely low and needs to be quickly raised to stabilize patient health [58].

Besides NO, which plays an important role in vasodilation, excess production of reactive oxygen species (ROS), including superoxide anion radicals ($O^{2-\bullet}$) and hydroperoxyl radicals ($HO_2^{\bullet}$), during illness or trauma can cause extreme oxidative stress and pose a significant threat to cells, tissues, and DNA. To neutralize the harmful effects of ROS, our body relies on a special class of metalloenzymes called superoxide dismutases (SODs). SODs play a pivotal role in controlling the levels of $O^{2-\bullet}$ in cells. They achieve this by catalytically converting $O^{2-\bullet}$ into O_2 and H_2O_2. The H_2O_2 generated by SOD is further catalytically transformed into O_2 and water by glutathione peroxidase or catalase [4]. There are three natural SODs: CuSOD, ZnSOD, and MnSOD, which are located intracellularly, extracellularly and in mitochondria, respectively [4]. These SODs are extremely crucial for our defense mechanism against superoxide radicals. However, in cases where the amount of superoxide radicals exceeds the limit that can be managed by natural SODs, SOD-mimicking drugs become increasingly important. These SOD mimics assist the body's defense mechanism.

Several metals like iron(II), copper(II), manganese(II), etc., derived macrocyclic complexes of porphyrins, phthalocyanines, porphyrazines, cyclic polyamines, and SALEN have been investigated as SOD mimics. Among them, the potential of manganese(II) macrocycles, particularly M40403, as SOD-mimicking agents holds great significance in mitigating the damaging effects of oxidative stress. The ability of MnSODs to neutralize superoxide species and provide a protective effect against ROS-related damage makes them a promising candidate for therapeutic applications. In addition to these features, M40403 also exhibit better stability and lower toxicity compared to copper and iron macrocycles [4].

Metallodrug for Gastrointestinal Disorder

Metallodrugs of sodium, magnesium, calcium, and aluminum, derived from carbonates, hydrogen carbonates, or hydroxide salts, are commonly used to alleviate minor stomach pain and digestion problems. These drugs neutralize excessive acidity in the stomach by raising the pH and reducing the secretion of acid by gastric cells thereby providing relief from symptoms like heartburn. Magnesium hydroxide, commonly known as 'Milk of Magnesia', also serves as a laxative to relieve constipation in addition to being an antacid. Besides the above-mentioned metal compounds, bismuth subsalicylate, also known as 'pink stuff', has been widely used to treat minor stomach ailments and alleviate symptoms of indigestion, diarrhea, heartburn, and nausea [4].

NUCLEAR MEDICINE

Nuclear medicine is a special medical procedure that utilizes radioactive formulations also known as radiopharmaceuticals, for medical use. Radiopharmaceuticals (RPs) can be categorized either on the basis of medical application, i.e., diagnostic or therapeutic, or by their bio-distribution. The bio-distribution of RPs is governed by their physical and chemical properties or by the receptor binding attached to the RPs. The latter class is often termed target-specific radiopharmaceuticals. Clinical applications of nuclear medicine were initiated with the use of ^{31}P and ^{131}I to treat leukemia and thyroid disorders, respectively. However, the real growth of the field was witnessed with the development of the ^{99}Mo-^{99m}Tc generator, when scientists discovered a way to produce the short-lived ^{99m}Tc from long-lived ^{99}Mo at Brookhaven National Laboratory, USA, in 1959 [59]. The wide spread use of nuclear medicine for diagnosis, management, and treatment of many serious diseases is a result of simultaneous advances in various fields, including the feasibility of producing radioisotopes in nuclear reactors, the synthesis of diverse molecular carriers for radiolabeling, and the development of sophisticated instruments for acquiring excellent quality images.

Nuclear medicine displays the unique ability to provide functional and physiological information, along with its diagnostic and therapeutic capabilities, which make it a useful modality in disease detection, management, and treatment [59]. One of the biggest advantages of nuclear medicine lies in its ability to detect abnormalities at an early stage, well before they can be detected by other diagnostic tools. Furthermore, RPs provide valuable insights into the biochemical and physiological functions of organs or tissues in real-time, which is hard to expect from other imaging modalities. Besides diagnostic applications, nuclear medicine could also be employed for therapeutic applications, wherein radioactive formulations are administered to target or destroy cancer cells and abnormal tissues. This mode of treatment is far more effective in cases where other modalities are insufficient or tumors are not accessible for surgery. Additionally, nuclear medicine offers an attractive avenue for theranostics, which combines diagnostic and therapeutic capabilities. This allows medical practitioners to accurately diagnose and monitor a disease while simultaneously delivering targeted therapy.

Design of Target Specific Radiopharmaceuticals

RPs are radiolabeled drugs that can be administered either orally or intravenously for diagnostic or therapeutic purposes. Since they involve the use of radioisotopes, adequate precautions must be taken for their safe use. Since the very beginning, the development of target-specific RPs has always been the aim in this field. Target-specific RPs play a key role in diagnostic and therapeutic applications, as they are capable of delivering the radiolabeled drug selectively to the target site. The basic design of target-specific RPs involves the association of a biomolecule with a radionuclide complex through a linker molecule. The biomolecule serves as a carrier, guiding the RP to the target site. Various biomolecules, such as monoclonal antibodies, antibody fragments, and small peptides, have been investigated as carrier molecules for radionuclides [59].

Linker molecules play a crucial role in target-specific RPs. Linkers not only act as spacers between the biomolecule and the radionuclide, thereby helping preserve the bio-affinity of the biomolecule, but they also contribute to improving the pharmacokinetics of the RP. However, some target-specific radio-formulations are known without the use of a linker molecule.

Additionally, bifunctional chelators (BFCs) could be used for the conjugation of the biomolecule at one end and the chelation of the radionuclide at the other end. The choice of BFC is primarily governed by the nature and oxidation state of the radionuclide [59]. It is generally preferred to position the radionuclide chelator moiety away from the receptor-binding motif to minimize possible interference between the two. The BFC approach is quite common in designing target-specific RPs, as it provides flexibility and control over the conjugation process. Various BFCs have been evaluated for RP formulation. Among them, BFCs based on polyaminopolycarboxylic ligands are quite favorable, as they form stable complexes with a wide variety of metal ions. The formation of stable complexes is crucial to avoid undesired in vivo release of metal ions. Macrocyclic-based BFCs facilitate the formation of thermodynamically and kinetically inert complexes.

Diagnostic Radiopharmaceuticals

Diagnostic RPs are radiolabeled molecules consisting of a radioisotope that either emits γ-rays for single-photon emission computed tomography (SPECT) or positrons for positron emission tomography (PET). Ideal diagnostic RPs should be pure γ or positron emitters of suitable energy and should not emit any particulate radiation. The emitted γ-photons should be high in abundance to obtain images of better resolution. Furthermore, it is crucial that the RP be produced with high specific activity and high radionuclide purity to minimize the risk of radiation exposure from impurity atoms. The radionuclide used for diagnostic purposes should have a sufficient half-life to allow the preparation of the radiochemical formulation and, at the same time, should be short enough to ensure the fading of radioactivity either through decay or excretion after the imaging procedure. Besides these aspects, high stability of RPs, ease of preparation, and the ability to selectively accumulate in the target organs are key factors for ideal diagnostic RPs. At this point, it is worth discussing the principles of the two most common diagnostic modalities, i.e., SPECT and PET.

Single Photon Emission Computed Tomography (SPECT)

SPECT relies on the detection of γ-photons emitted from radiopharmaceuticals administered to the patient. The most commonly used radioisotope for SPECT is technetium-99m, which has a half-life of 6.02 hours. Apart from this, other radionuclides employed in SPECT include ^{67}Ga (3.26 days), ^{111}In (2.81 days), and ^{201}Tl (72.91 hours). In SPECT measurements, an RP, usually attached to a carrier biomolecule, is administered intravenously. Upon radioactive decay, it emits γ-photons. The γ cameras, such as scintillation-based NaI:Tl detectors are equipped with lead collimators to allow the detection of γ-photons emitted from a specific direction. The detector converts the incoming γ-photons into corresponding electrical signals proportional to the intensity of the detected γ-rays. These electrical signals are processed using computational algorithms to generate a 3-dimensional image corresponding to the distribution of the RP within the patient's body. Instead of just reconstructing the anatomical image of the patient, SPECT provides valuable information about biological activity at each location. The emission of the radionuclide indicates the amount of blood flow in the capillaries of the region under observation. Thus, SPECT measurement allows accurate localization in 3D space, thereby providing valuable information about the biological activity of internal organs, including brain and cardiac imaging. Various radiochemical formulations, such as ^{111}In-DTPA (diethylenetriaminepentaacetic acid) and ^{99m}Tc-HYNIC-TATE (6-hydrazinopyridine-3-carboxylic acid derivatized Tyr^3, Thr^8-octreotide), have been used as SPECT RPs for diagnosing somatostatin receptor-positive neuroendocrine tumors. It is worth nothing at this point that neuroendocrine tumors overexpress somatostatin receptors. After extensive investigation, octreotide has been developed, which mimics somatostatin and therefore guides the RP to selectively accumulate in regions where there is overexpression of somatostatin receptors [60].

Positron Emission Tomography (PET)

PET is a sophisticated non-invasive diagnostic tool that uses a radiopharmaceutical containing a positron-emitting isotope to investigate the metabolic activity of cells and tissues. It is widely used in clinical oncology, cardiology, neurology, and in the diagnosis, staging, and monitoring of various diseases, particularly cancer. By measuring metabolic activities within the cells and tissues, PET provides valuable information about the physiology and anatomy of the organ under evaluation [61].

PET relies on the measurement of two coincident γ-photons emitted due to positron-electron annihilation. Unlike SPECT, PET does not require collimators because, in PET scans, coincidence measurements of γ-photon, take care of collimation [62]. Additionally, PET scans are known for their high sensitivity compared to SPECT. In PET scans, ~1% of emitted photons are measured compared to ~0.01% in the case of SPECT scans. This enhanced sensitivity allows PET to provide more accurate and detailed information about the physiological and structural changes occurring within the organ with high spatial resolution. It is also important to note that, whereas PET directly detects metabolism and visualizes the structural and functional changes within tissues, other nuclear diagnostic procedures, such as SPECT, measure the amount of radioactive substance in a particular location to extract information about tissue function. This key distinction allows PET to provide unique physiological insights within the organ, offering an overall understanding of the disease. Some important metal radionuclides used in PET imaging are ^{68}Ga, with a half-life of 68 minutes; ^{64}Cu, with a half-life of 12.7 hours; and ^{89}Zr, with a half-life of 78.4 hours. Among several radioformulations developed for PET, ^{68}Ga-DOTA-TATE (tetraazacyclododecane tetraacetic acid) has been approved by the FDA for imaging neuroendocrine tumors overexpressing somatostatin receptors [63,64].

Therapeutic Radiopharmaceuticals

These are radiolabeled molecules designed to deliver therapeutic doses of ionizing radiation to the target organ/site. The therapeutic effect of radiation can be exerted on the desired site in three ways, known as brachytherapy, teletherapy, and radionuclide therapy. Brachytherapy is a therapeutic modality wherein a sealed radiation source, called seeds, is implanted inside or just in the vicinity of the target site. It is widely used to treat skin, prostate, eye, and cervical cancer. The metal radionuclides commonly administered in brachytherapy are ^{137}Cs, ^{192}Ir and ^{106}Ru. It is important to note that brachytherapy is useful for the treatment of accessible tumors [65].

The second modality is external beam irradiation therapy, also commonly known as teletherapy. It is a widely used radiation therapy for the treatment of cancer, wherein ionizing radiations are directed at the tumor site from outside the body to kill cancer cells and shrink the tumor. The most commonly used radioisotope in teletherapy is ^{60}Co.

Radionuclide therapy involves the systematic administration of radiopharmaceuticals designed to selectively accumulate at the diseased site. In this modality, particulate emitter radioisotopes (e.g., α, β, and conversion electron) are used to impart therapeutic effects. Ideal radionuclide therapy involves the selective accumulation of radionuclides at the target site in sufficient concentration to deliver a cytotoxic radiation dose to the tumor site. Furthermore, the radionuclide, after imparting its therapeutic effect, should clear rapidly from the blood and other organs to reduce radiation exposure to normal tissues. Another important aspect is the half-life of the radionuclide, which should be long enough to allow its accumulation at the target site without significant radioactive decay. Besides these factors, high radionuclidic purity, high stability, and ease of sample preparation are crucial factors to consider when preparing radioformulations for therapeutic purposes [65].

Various radionuclides have been used for the preparation of suitable RPs. Among them, $^{89}SrCl_2$ (β– emitter; half-life 50.53 days), ^{153}Sm-EDTMP (β^- and γ emitter; half-life 46.27 hours), ^{177}Lu-EDTMP (β^- and γ emitter; half-life 6.73 days), and ^{188}Re-HEDP (β^- and γ emitter; half-life 90.64 hours) (EDTMP=ethylene diaminetetramethylene phosphonate; HEDP=hydroxyethylenediphosphonate) are approved agents for metastatic bone pain palliation therapy. ^{90}Y-citrate (β^- emitter; half-life 60.10 hours) is particularly used for rheumatoid arthritis, and ^{90} Y-solid glass microspheres are used for hepatocellular carcinoma treatment. Peptide-modified RPs have been successfully used for targeted therapy. Some notable examples include ^{177}Lu-DOTA-TATE and ^{225}Ac-DOTA-TATE (α and γ emitter; half-life 10 days), which are widely used to treat neuroendocrine cancers, while ^{177}Lu-PSMA and ^{225}Ac-PSMA (PSMA=prostate-specific membrane antigen) are used for the treatment of prostate cancer [66].

Radiopharmaceuticals can be utilized for theranostic purposes, meaning they can simultaneously serve both therapeutic and diagnostic roles. The term 'theranostics' was originally coined by John Funkhouser in 1998, emphasizing personalized treatment. Theranostics involves a comprehensive and customized pharmacotherapy approach that improves therapeutic effects while minimizing treatment toxicities. In nuclear medicine, theranostics involves pairing therapeutic and diagnostic radioactive agents, chelated to a carrier molecule, and targeted toward specific clinical conditions. Therapeutic radioisotopes decay by releasing particles like α, β-, or auger electrons, capable of ionization or bond breakage, while diagnostic radioisotopes decay by releasing gamma rays or emit gamma rays after positron annihilation, which are used for imaging purposes. Ideally, a theranostic pair in nuclear medicine consists of radioisotopes derived from the same element, one for therapeutic action and the other for diagnosis. However, such conjugations are very rare. There is another concept called 'twins in spirit' [67]. This term describes a pair of compounds that may or may not be identical in their chemical or biological composition. However, what makes them akin in spirit is the unique ability of the diagnostic counterpart to accurately predict the biodistribution of the therapeutic radionuclide. One such pair is ^{68}Ga-^{177}Lu, which has achieved remarkable success and is routinely employed in the treatment of neuroendocrine tumors (NET). The two isotopes play complementary roles in this theranostic approach. ^{68}Ga (half-life: 68 min) serves as a common choice for PET imaging and can be easily obtained from a ^{68}Ge/^{68}Ga generator system. On the other hand, ^{177}Lu (half-life: 6.7 days) is a beta-emitter. Thus, in the ^{68}Ga-^{177}Lu theranostic pair, ^{68}Ga is used for imaging, while ^{177}Lu is utilized for radiotherapy [67].

Overall, the utilization of radiation-emitting isotopes for diagnostic, therapeutic and theranostic purposes has revolutionized the medical field. On one hand, they enable the non-invasive diagnosis of diseases with high spatial resolution, and on the other hand, they offer excellent therapeutic avenues for target-specific treatment.

ROLE OF NANOMATERIALS IN THERAPEUTICS AND DIAGNOSTICS

Over time, nanomaterials have emerged as excellent tools in both therapeutics and diagnosis. Their significant success in the medical field is attributed to their unique characteristics, including a large surface area-to-mass ratio, high surface reactivity, tunable physiochemical properties, and intriguing optical, electrical and magnetic properties. Importantly, nanomaterials can be engineered to suit particular applications by controlling their size and morphology—from quantum dots (0D) to nanorods (1D) to nanosheets (2D) to nanoprisms and nanoflowers (3D)—as well as their composition and surface modifications. These appealing features of nanomaterials have been exploited to develop state-of-the-art diagnostic and therapeutic platforms to serve present as well as future needs.

The importance of nanomaterials lies in their ability to develop accurate and sensitive point-of-care (PoC) diagnostic tools, i.e., devices that can perform diagnostic tests near the patient. For the development of PoCs, nanomaterials are conjugated to a biomarker to detect the presence of a given analyte or pathogen. These functionalized nanoparticles have a wide range of detection mechanisms depending on the nature of the nanoparticles [64]. The most widely used detection mechanism involves the measurement of absorption or emission features of nanoparticles, including surface plasmon resonance arising due to the disturbance of the dielectric constant induced by adsorption of a molecule, a shift in the peak wavelength due to agglomeration, or quenching or enhancement of emission intensity arising due to enzymatic reactions on the surface of the nanoparticle. Additionally, electrical or electrochemical changes that occur upon interaction with a particular analyte also serve as an attractive detection mechanism. The most widely used nanoparticles for developing sensitive PoC biosensors are carbon quantum dots, graphene and graphene-oxide nanosheets, carbon nanotubes, gold nanoparticles (AuNPs), and cadmium selenide quantum dots [68].

Magnetic nanomaterials, such as magnetic iron oxide nanoparticles, have attracted enormous attention as contrast agents for magnetic resonance imaging and serve as an alternative to gadolinium chelates, first-generation MRI contrast agents [69]. The use of gadolinium chelates has been associated with the possibility of nephrotoxicity.

Nanomaterials possess unique capabilities that enable them to penetrate and accumulate at biological sites through the enhanced permeability and retention effect. They can swiftly cross physiological barriers and possess specific recognition and binding to target areas. These extraordinary features have ignited significant interest in developing nanomaterial-based therapeutics that aim to overcome the limitations often faced in conventional therapeutics.

Nanomaterials have been widely used as carriers to deliver photosensitizers to target sites for photodynamic therapy [70]. AuNPs functionalized with photosensitizers and biomolecules have been explored to target cancer cell surfaces.

Some nanomaterials, such as copper sulfide ($Cu_{2-x}S$), exhibit photothermal conversion properties. This unique characteristic of nanoparticles to convert light into heat is extensively exploited to selectively ablate tumor cells. Furthermore, nanoparticles can also be tuned to exhibit both photodynamic and photothermal activity simultaneously, allowing for dual-mode treatment [71]. Magnetic nanoparticles (MNPs) are also used in magnetic hyperthermia for cancer therapy. This primarily non-invasive cancer therapy involves the delivery and selective accumulation of MNPs at the tumor site. These MNPs are then activated by an alternating magnetic field to generate thermal energy through hysteresis losses. The aim of this technique is to selectively raise the temperature of the tumor region to 45°C–46°C. Sufficient exposure of the tumor region to this temperature is enough to cause thermoablation [72]. This modality is particularly useful for deep-seated tumors that are difficult to remove through surgery. Fe_2O_3 nanoparticles are the most widely explored MNPs for magnetic hyperthermia.

Besides these applications, nanoparticles have emerged as promising drug delivery vehicles. Surface functionalization of drug-conjugated nanoparticles with targeting ligands, such as antibodies or peptides, enables drug delivery to the target site. Engineered nanoparticles exhibit an extraordinary ability to bypass biological barriers. This characteristic offers the advantage of enhanced drug delivery efficiency. Additionally, nanoparticles coated with drugs exhibit enhanced stability in biological media. Nanoparticles as drug carriers allow for controlled and sustained release of drug molecules at the target site upon mild triggers [73]. Furthermore, nanoparticle-conjugated drugs could exhibit improved pharmacokinetics in terms of absorption, distribution, and excretion. Drugs conjugated to functionalized nanoparticles exhibiting fluorescence or magnetic resonance, among other features, could be used for dual purposes i.e., drug delivery and real-time monitoring of therapeutic response. Overall, nanoparticles have exhibited tremendous potential in a wide range of biomedical applications, including diagnosis and therapeutics.

CONCLUSION

The chapter emphasizes the crucial and diverse roles that metal ions play in the fields of therapeutics and diagnostics. The choice of metal, ligands, and coordination geometry contributes to their unique mechanisms of action. The versatility of metallodrugs is demonstrated in their potential as anticancer, antibacterial, antiviral, antiparasitic, and anti-inflammatory agents, among others. The concept of using metallodrugs as prodrugs and the methods of activation has been elucidated with the help of Pt(IV) prodrugs. Furthermore, the importance of radiopharmaceuticals as valuable diagnostic, therapeutic, and theranostic tools is underscored. Finally, the chapter explores the promising capabilities of nanomaterials in point-of-care diagnostics and as effective therapeutic agents. This comprehensive exploration is highly useful for understanding the vital role of metal ions in medical applications.

REFERENCES

1. C. Orvig, M. J. Abrams, Medicinal inorganic chemistry: introduction, *Chem. Rev.* 1999, 99, 2201–2204.
2. M. Zhao, Y. Li, Z. Wang, Mercury and mercury-containing preparations: history of use, clinical applications, pharmacology, toxicology, and pharmacokinetics in traditional Chinese medicine, *Front Pharmacol.* 2022, 13, 807807.
3. K. J. Williams, The introduction of 'chemotherapy'using arsphenamine–the first magic bullet, *J. R. Soc. Med.* 2009, 102, 343–348.
4. K. D. Mjos, C. Orgiv, Metallodrugs in medicinal inorganic chemistry, *Chem. Rev.* 2014, 114, 4540–4563.
5. N. C. Lloyd, H. W. Morgan, B. K. Nicholson, R. S. Ronimus, The composition of Ehrlich's Salvarsan: resolution of a century-old debate, *Angew. Chem. Int. Ed.* 2005, 44, 941.
6. T. Lazarevic, A. Rilak, Z. D. Bugarcic, Platinum, palladium, gold and ruthenium complexes as anticancer agents: Current clinical uses, cytotoxicity studies and future perspectives, *Eur. J. Med. Chem.* 2017, 142, 8–31.
7. S. J. B. Price, A. Filipovska, Gold compounds as therapeutic agents for human diseases, *Metallomics* 2011, 3, 863–873.
8. K. J. Franz, N. M. Nolte, Introduction: metals in medicine, *Chem. Rev.* 2019, 119, 727–729.
9. N. P. E. Barry, P. J. Sadler, Exploration of the medical periodic table: towards new targets, *Chem. Commun.* 2013, 49, 5106–5131.
10. E. J. Anthony, E. M. Bolitho, H. E. Bridgewater, O. W. L. Carter, J. M. Donnelly, C. Imberti, E. C. Lant, F. Lermyte, R. J. Needham, M. Palau, P. J. Sadler, H. Shi, F. –X. Wang, W.-Y. Zhang, Z. Zhang, Metallodrugs are unique: opportunities and challenges of discovery and development, *Chem. Sci.* 2020, 11, 12888–12917.
11. C. Sissi, M. Palumbo, Effects of magnesium and related divalent metal ions in topoisomerase structure and function, *Nucleic Acids Res.* 2009, 37, 702–711.
12. E. Meggers, Exploring biologically relevant chemical space with metal complexes, *Curr. Opin. Chem. Biol.* 2007, 11, 287–292.
13. A. Frei, J. Zuegg, A. G. Elliott, M. Baker, S. Braese, C. Brown, F. Chen, C. G. Dowson, G. Dujardin, N. Jung, A. P. King, A. M. Mansour, M. Massi, J. Moat, H. A. Mohamed, A. K. Renfrew, P. J. Rutledge, P. J. Sadler, M. H. Todd, C. E. Willans, J. J. Wilson, M. A. Cooper, M. A. T. Blaskovich, Metal complexes as a promising source for new antibiotics, *Chem. Sci.* 2020, 11, 2627–2639.
14. C. A. Lipinski, F. Lombardo, B. W. Dominy, P. J. Feeney, Experimental and computational approaches to estimate solubility and permeability in drug discovery and development settings, *Adv. Drug Deliv. Rev.* 2001, 46, 3–26.
15. E. R. Jamieson, S. J. Lippard, Structure, recognition, and processing of Cisplatin-DNA adducts, *Chem Rev.* 1999, 8, 2467–2498.
16. S. G. Chaney, S. L. Campbell, B. Temple, E. Bassett, Y. Wu, M. Faldu, Protein interactions with platinum–DNA adducts: from structure to function, *J. Inorg. Biochem.* 2004, 10, 1551–1559.
17. J. W. Cox, S. J. Berners-Price, M. S. Davies, Y. Qu, N. Farrell, Kinetic analysis of the stepwise formation of a long-range DNA interstrand cross-link by a dinuclear platinum antitumor complex: evidence for aquated intermediates and formation of both kinetically and thermodynamically controlled conformers, *J. Am. Chem. Soc.* 2001, 123, 1316–1326.
18. H.-K. Liu, P. J. Sadler, Metal complexes as DNA intercalators, *Acc. Chem. Res.* 2011, 44, 349–359.

19. M. V. Keck, S. J. Lippard, Unwinding of supercoiled DNA by platinum-ethidium and related complexes, *J. Am. Chem. Soc.* 1992, 114, 3386–3390.
20. N. J. Wheate, R. I. Taleb, A. M. Krause-Heuer, R. L. Cook, S. Wang, V. J. Higgins, J. R. Aldrich-Wright, Novel platinum (II)-based anticancer complexes and molecular hosts as their drug delivery vehicles, *Dalton Trans.* 2007, 5055–5064.
21. T. R. Steel, C. G. Hartinger, Metalloproteomics for molecular target identification of protein-binding anticancer metallodrugs, *Metallomics* 2020, 12, 1627–1636.
22. D. Criscuolo, R. Avolio, D. S. Matassa, F. Esposito, *Front. Onco.* 2021, 11, https://doi.org/10.3389/fonc.2021.797265.
23. J. Maksimoska, L. Feng, K. Harms, C. Yi, J. Kissil, R. Marmorstein, E. Meggers, Targeting large kinase active site with rigid, bulky octahedral ruthenium complexes, *J. Am. Chem. Soc.* 2008, 130, 15764–15765.
24. A. Y. Chen, R. N. Adamek, B. L. Dick, C. V. Credille, C. N. Morrison, S. M. Cohen, Targeting metalloenzymes for therapeutic intervention, *Chem. Rev.* 2019, 119, 1323–1455.
25. R. Ye, C. Tan, B. Chen, R. Li, Z. Mao, Zinc-containing metalloenzymes: inhibition by metal-based anticancer agents, *Front. Chem.* 2020, 8, 402.
26. T.C. Johnstone, K. Suntharalingam, S. J. Lippard, The next generation of platinum drugs: targeted Pt (II) agents, nanoparticle delivery, and Pt (IV) prodrugs, *Chem. Rev.* 2016, 116, 3436–3486.
27. A. F. A. Peacock, A. Habtemariam, R. Fern´andez, V. Walland, F. P. A. Fabbiani, S. Parsons, R. E. Aird, D. I. Jodrell, P. J. Sadler, Tuning the reactivity of osmium (II) and ruthenium (II) arene complexes under physiological conditions, *J. Am. Chem. Soc.* 2006, 128, 1739–1748.
28. Z. Liu, I. Romero-Canel´on, B. Qamar, J. M. Hearn, A. Habtemariam, N. P. E. Barry, A. M. Pizarro, G. J. Clarkson, P. J. Sadler, The potent oxidant anticancer activity of organoiridium catalysts, *Angew. Chem., Int. Ed.* 2014, 53, 3941–3946.
29. X. Wang, X. Wang, S. Jin, N. Muhammad, Z. Guo, Stimuli-responsive therapeutic metallodrugs, *Chem. Rev.* 2019, 119, 1138–1192.
30. X. Zhang, X. Li, Q. You, X. Zhang, Prodrug strategy for cancer cell-specific targeting: a recent overview, *Eur. J. Med. Chem.* 2017, 139, 542–563.
31. I. R. Canelon, P. J. Sadler, Next-generation metal anticancer complexes: multitargeting via redox modulation, *Inorg. Chem.* 2013, 52, 12276–12291.
32. M. D. Hall, H. R. Mellor, R. Callaghan, T. W. Hambley, Platinum (IV) antitumour compounds: their bioinorganic chemistry, *Coord. Chem. Rev.* 2002, 232, 49–67.
33. E. Amtmann, M. Zoller, H. Wesch, G. Schilling, Antitumoral activity of a sulphur-containing platinum complex with an acidic pH optimum, *Cancer Chemother. Pharmacol.* 2001, 47, 461–466.
34. H.S. Oberoi, F. C. Laquer, L. A. Marky, A. V. Kabanov, T. K. Bronich, Core cross-linked block ionomer micelles as pH-responsive carriers for cis-diamminedichloroplatinum (II), *J. Contol Release* 2011, 153, 64–72.
35. S. Ding, A. J. Pickard, G. L. Kucera, U. Bierbach, Design of enzymatically cleavable prodrugs of a potent platinum-containing anticancer agent, *Chem. Eur. J.* 2014, 20, 16164–16173.
36. C. Mari, V. Pierroz, S. Ferrari, G. Gasser, Combination of Ru (II) complexes and light: new frontiers in cancer therapy, *Chem. Sci.* 2015, 6, 2660–2686.
37. S. Bonnet, Why develop photoactivated chemotherapy? *Dalton Trans.* 2018, 47, 10330–10343.
38. A. Presa, G. V´azquez, L. A. Barrios, O. Roubeau, L. Korrodi-Greg´orio, R. P´erez-Tom´as, P. Gamez, Photoactivation of the cytotoxic properties of platinum (II) complexes through ligand photoswitching, *Inorg. Chem.* 2018, 57, 4009–4022.

39. Y. Liu, P. Bhattarai, Z. Dai, X. Chen, Photothermal therapy and photoacoustic imaging via nanotheranostics in fighting cancer, *Chem. Soc. Rev.* 2019, 48, 2053–2108.
40. F. Liu, V. Kozlovskaya, S. Medipelli, B. Xue, F. Ahmad, M. Saeed, D. Cropek, E. Kharlampieva, Temperature-sensitive polymersomes for controlled delivery of anticancer drugs, *Chem. Mater.* 2015, 27, 7945–7956.
41. S.-C. Song, S. B. Lee, B. H. Lee, H.-W. Ha, K.-T. Lee, Y. S. Sohan, Synthesis and antitumor activity of novel thermosensitive platinum (II)–cyclotriphosphazene conjugates, *J. Controlled Release* 2003, 90, 303–311.
42. M. R. Gill, K. A. Vallis, Transition metal compounds as cancer radiosensitizers, *Chem. Soc. Rev.* 2019, 48, 540–557.
43. T.C. Johnstone, J. J. Wilson, S. J. Lippard, Monofunctional and higher-valent platinum anticancer agents, *Inorg. Chem.* 2013, 52, 12234–12249.
44. S. J. Lippard, Metal in medicine. In I. Bertini, H.B. Gray, S.J. Lippard, J.S. Valentine (Eds.), *Bio-Inorganic Chemistry*, University Science Books, Mill Valley, CA, 1994, page 505.
45. I. Yousuf, M. Bashir, F. Arjmand, S. Tabassum, Advancement of metal compounds as therapeutic and diagnostic metallodrugs: Current frontiers and future perspectives, *Coord. Chem. Rev.* 2021, 445, 214104.
46. R. Crichton, Chapter 22, Metals in Medicine and Metal in Drugs. In *Biological Inorganic Chemistry*, Academic Press, Elsevier, Amsterdam, The Netherlands, 2019.
47. E. Jortzik, M. Farhadi, R. Ahmadi, Antiglioma activity of GoPI-sugar, a novel gold (I)–phosphole inhibitor: Chemical synthesis, mechanistic studies, and effectiveness in vivo, *Biochim. Biophys. Acta* 2014, 1844, 1415–1426.
48. C. R. Chitambar, Gallium and its competing roles with iron in biological systems, *Pharmacol. Res.*, 2016, 115, 56–64.
49. Microbion Corp., Bozeman (MT); https://www.microbioncorp.com.
50. E. Kabir, M. R. O. K. Noyon, M. A. Hossain, Synthesis, biological and medicinal impacts of metallodrugs: a studym, *Results in Chem.* 2023, 5, 100935.
51. J. A. Schwartz, E. K. Lium, S. J. Silverstein, Herpes simplex virus type 1 entry is inhibited by the cobalt chelate complex CTC-96, *J. Virol.* 2001, 75, 4117–4128.
52. J. Wang, Y. Liu, K. Xu, Y. Qi, J. Zhong, K. Zhang, J. Li, E. Wang, Z. Wu, Z. Kang, Broad-spectrum antiviral property of polyoxometalate localized on a cell surface, *ACS Appl. Mater. Interfaces* 2014, 6, 9785–9789.
53. N. Yang, J. A. Tanner, B. J. Zheng, R. M. Watt, M. L. He, L. Y. Lu, J. Q. Jiang, K. T. Shum, Y. P. Lin, K. L. Wong, M. C. Lin, H. F. Kung, H. Sun, J. D. Huang, Bismuth complexes inhibit the SARS coronavirus, *Angew. Chem., Int. Ed.*, 2007, 46, 6464–6468.
54. H. A. Rothan, S. Stone, J. Natekar, P. Kumari, K. Arora, M. Kumar, The FDA-approved gold drug auranofin inhibits novel coronavirus (SARS-COV-2) replication and attenuates inflammation in human cells, *Virology* 2020, 547, 7–11.
55. A. J. te Velthuis, S. H. van den Worm, A. C. Sims, R. S. Baric, E. J. Snijder, M. J. van Hemert, Zn2+ inhibits coronavirus and arterivirus RNA polymerase activity in vitro and zinc ionophores block the replication of these viruses in cell culture, *PLoS Pathog.* 2010, 6, e1001176.
56. H. HogenEsch, D. T. O'Hagan, C. B. Fox, Optimizing the utilization of aluminum adjuvants in vaccines: you might just get what you want, *npj Vaccines*, 2018, 3, 1–11.
57. L. H. Price, G. R. Heninger, Lithium in the treatment of mood disorders, *N. Engl. J. Med.*, 1994, 331, 591.
58. B. R. Cameron, M. C. Darkes, I. R. Baird, R. T. Skerlj, Z. L. Santucci, S. P. Fricker, Ruthenium (III) triazacyclononane dithiocarbamate, pyridinecarboxylate, or aminocarboxylate complexes as scavengers of nitric oxide, *Inorg. Chem.*, 2003, 42, 4102.

59. S. Liu, The role of coordination chemistry in the development of target-specific radiopharmaceuticals, *Chem. Soc. Rev.* 2004, 33, 445–461.
60. A. Mukherjee, A. Korde, A. Shinto, H. D. Sarma, K. Kamaleswaran, A. Dash, Studies on batch formulation of a freeze dried kit for the preparation of 99mTc-HYNIC-TATE for imaging neuroendocrine tumors, *Appl. Radiat. Isot.* 2019, 145, 180–186.
61. https://www.hopkinsmedicine.org/health/treatment-tests-and-therapies/positron-emission-tomography-pet.
62. S. Walrand, M. Hesse, F. Jamar, Update on novel trends in PET/CT technology and its clinical applications, *Br. J. Radiol.* 2018, 91, 20160534.
63. M. S. Hofman, W. F. E. Lau, R. J. Hicks, *RadioGraphics* 2015, 35, https://doi.org/10.1148/rg.352140164
64. M. Perfe´zou, A. Turner, A. Merkoçi, Cancer detection using nanoparticle-based sensors, *Chem. Soc. Rev.* 2012, 41, 2606–2622.
65. G. Sgouros, L. Bodei, M. R. McDevitt, J. R. Nedrow, Radiopharmaceutical therapy in cancer: clinical advances and challenges, *Nat. Rev. Drug Discov.* 2020, 19, 589–608.
66. N. K. Tafreshi, D. N. Pandya, C. J. Tichacek, M. M. Budzevich, Z. Wang, J. N. Reff, R. W. Engelman, D. C. Boulware, A. A. Chiappori, J. R. Strosberg, H. Ji, T. J. Wadas, G. E. Haddad, D. L. Morse, Preclinical evaluation of [225 Ac] Ac-DOTA-TATE for treatment of lung neuroendocrine neoplasms, *Eur. J. Nuc. Med. Mol. Imaging* 2021, 48, 3408–3421.
67. N. Naskar, S. Lahiri, Theranostic terbium radioisotopes: challenges in production for clinical application, *F. Med.* 2021, 8, 675014.
68. D. Q. Gonzalez, A. Merkoci, Nanomaterial-based devices for point-of-care diagnostic applications, *Chem. Soc. Rev.* 2018, 47, 4697–4709.
69. Z. Shen, A. Wu, X. Chen, Iron oxide nanoparticle based contrast agents for magnetic resonance imaging, *Mol. Pharmaceutics* 2017, 14, 1352–1364.
70. P. G. Calavia, G. Bruce, L. P. Gracia, D. A. Russell, Photosensitiser-gold nanoparticle conjugates for photodynamic therapy of cancer, *Photochem. Photobiol. Sci.* 2018, 17, 1534–1552.
71. L. Han, Y. Zhang, X. –W. Chen, Y. Shu, J.-H Wang, Protein-modified hollow copper sulfide nanoparticles carrying indocyanine green for photothermal and photodynamic therapy, *J. Mater. Chem. B* 2016, 4, 105–112.
72. C. L. Dennis, R. Ivkov, Physics of heat generation using magnetic nanoparticles for hyperthermia, *Int. J. Hypethermia* 2013, 29, 715–729.
73. X. Yang, M. Yang, B. Pang, M. Vara, Y. Xia, Gold nanomaterials at work in biomedicine, *Chem. Rev.* 2015, 115, 10410–10488.

10 Biomineralization

INTRODUCTION

Bioinorganic solid-state chemistry, which is usually referred to as the field of biomineralization, covers the structure and synthesis of inorganic materials or organic-inorganic hybrids in a biological environment. The minerals obtained through the biomineralization process are called biominerals. The 'bio' in the term biomineral indicates that the high activation energy required for the preparation of inorganic materials through conventional solid-state synthesis routes is bypassed by the involvement of biomolecules in the synthesis. Living organisms play a crucial role in constructing a wide variety of organic/inorganic hybrid materials, called biominerals, through the self-organization of organic molecules and inorganic elements under ambient conditions. Biominerals usually have highly organized and hierarchical structures ranging from nanometer to micrometer length scales. Additionally, being synthesized under a strictly organism-controlled environment, biominerals possess shape, size, crystallinity, trace element composition, and other properties that are impossible to observe in their inorganically formed counterparts [1]. Biomineralization has historical roots coinciding with the first evidence of life is and often regarded as an important factor behind nature's evolutionary success. The first evidence of biomineralization is safeguarded in the form of fossil stromatolites (bacteria-like prokaryotic organisms) found in sedimentary rocks. Gradually, organisms improved their ability to control mineral formation and have produced a stunning array of biomaterials with functional roles from limited available resources through extremely complex reactions operating under highly regulated conditions. Today, the phenomenon of biomineralization is widespread; members of all five kingdoms of organisms are able to produce minerals. The mechanical, optical, and magnetic properties of these materials are exploited by organisms for a variety of purposes [2]. These properties are often optimized for a given function and are different from the properties of other biological materials of similar composition. Materials chemists are intrigued by the exceptional control organisms exert over the composition, crystallography, morphology, and material properties of biominerals, as well as the mild conditions (physiological temperature, pressure, and pH) required to prepare them. Frequently cited examples include the formation of bones and shells, which provide key fossil records that yield important insights into history, archaeology, and anthropology. Other illustrative examples include hydroxyapatite in the bones and teeth of mammals, calcium carbonate in molluscan shells, calcite single crystals found in the skeletal construction of brittle stars, silica in diatoms and marine sponges, and magnetite in magnetotactic bacteria. These biominerals have elaborate hierarchical structures, giving them intriguing properties such as

DOI: 10.1201/9781003432333-10

high mechanical hardness, flexibility, and functionality, which are not provided by conventional synthetic materials [3–5]. For instance, magnetite nanocrystals (Fe_3O_4) in organisms like birds and bacteria function as magnetic sensors that help them in navigation. Calcite single crystals in the skeletons of brittle stars function as photosensory organs. Apart from performing functional roles, biominerals also provide tensile strength that holds the skeleton of organisms together. In short, biominerals perform diverse functions such as tensile support, protection, sensing, storage, and even homeostasis.

BIOMINERALIZATION VS INORGANIC SYNTHESIS

Biominerals differ significantly from inorganically (laboratory/geogenic) synthesized minerals. Firstly, biominerals exhibit unique morphologies (Figure 10.1) and functional properties quite different from their inorganically produced counterparts. For example, molluscan shells, made up of calcium carbonate, have a fracture tolerance 3,000 times greater than inorganic crystals. The improved mechanical strength of the shell is attributed to the organic components occluded within the calcium carbonate crystals, and this marks the second important difference between biominerals and inorganic counterparts. Biominerals are usually composites or hybrid materials consisting of both inorganic and organic parts. In biominerals, inorganic crystals are either distributed within a complex macromolecular framework like collagen fibers in the case of hydroxyapatite, or in a mosaic domain delimited by organic layers or they may also contain organic macromolecules embedded within them.

One important characteristic of biomineralization is that it is a green process. In order to produce complex inorganic structures, as shown in Figure 10.1, organisms use aqueous solutions at temperatures well below the boiling point of water and generate no toxic intermediates [3,4]. From the perspective of rapidly growing global environmental issues, biomineralization-inspired synthetic processes

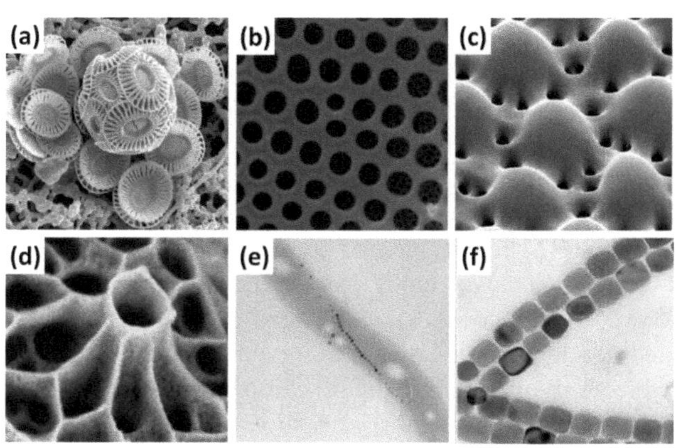

FIGURE 10.1 Representation of nature's ability to create staggering diversity of structures. (Copyright permission from American Chemical Society, Refs. [3–5].)

are of great importance in materials science to reduce energy consumption and promote environmental benignity [5].

In recent years, therefore, the field of biomineralization has expanded to include the application of strategies adapted from biology to the production of synthetic materials. Biomineralization is, by definition, a multidisciplinary field that draws on researchers from biology, chemistry, geology, materials science, and beyond.

MAJOR ELEMENTS INVOLVED IN BIOMINERALIZATION

Of the ~25 essential elements required by living organisms, calcium is probably the most widely distributed element in biominerals, especially in parts requiring high mechanical strength, like bones and teeth. Other metal ions involved include Si, Mg, Fe, Mn, Cu, Na, Ni, and Zn. These metal ions precipitate as hydroxides (OH), oxides (O^{2-}), sulfides (S^{2-}), carbonates ($CO_3^{2-}CO_3^{2-}$) and phosphates (PO_4^{3-} PO_4^{3-}). Some non-essential metal ions are either found in association with external cell walls of bacteria (Ag, Au, Pb, and U) or accumulated and deposited as intracellular minerals (Ba and Sr). Table 10.1 provides a list of important biominerals present in organisms and their functions.

The selection of metal ions for biominerals is based on the rule of bio-selection [6]. The first is *the rule of chemical fitness,* i.e., the element that is chemically fit for a certain biological role is selected. The second is *the rule of abundance*, which states that organisms prefer more abundant and readily available elements. These rules have been discussed at length in Chapter 1 of this book.

Among all the metal ions, calcium-containing biominerals are the most abundant, both in terms of the quantities produced and their prevalence among many different taxa [2]. Calcium-containing biominerals are generally used for skeleton formation, mechanical support, and protection. Occasionally, they are used for optical imaging and gravity receptors. The prevalence of calcium biominerals compared to other Group II metals is primarily because of the low solubility products of carbonates, phosphates, pyrophosphates, oxalates, sulfates, and relatively high levels of calcium in extracellular fluids. Compared to calcium, magnesium salts are generally more soluble, while the concentrations of strontium and barium are relatively lower. However, magnesium has a pronounced effect in influencing the structure of both carbonate and phosphate biominerals through lattice and surface substitution reactions [7]. Calcium prefers to bind with oxygen donor anions such as phosphates and carboxylates. Phosphates and carbonates of calcium are major components of bones, teeth, and shells of many organisms, such as mollusks, along with a complex organic macromolecular matrix consisting of proteins, polysaccharides, and lipids. Interestingly, calcium-containing biominerals are found to have high levels of acidic amino acids, like aspartate and glutamate residues [6].

A few unicellular organisms, like diatoms, use amorphous silica instead of calcium carbonate to form their outer coating, despite the fact that silicon is much less abundant than calcium and strontium in seawater [6]. This is probably due to the stability of Si-O-Si linkages and the fact that silica forms a transparent

TABLE 10.1
Special Function of Biominerals in Organism

Biomineral	Specialized Functions in Organism
Calcite ($CaCO_3$)	Coccolithophorids, Foraminifera, Mollusks: Exoskeletons
	Trilobites (eye lens): optical imaging
	Aves (egg shell): protection
	Mammalia (inner ear): gravity receptor
Aragonite ($CaCO_3$)	Scleractinian corals (cell wall): Exoskeletons
	Cephalopoda (shell): Buoyancy device
	Gastropoda: reproduction
	Fish (head): gravity receptor
Vaterite ($CaCO_3$)	Ascidians (spicules): protection
Amorphous ($CaCO_3$)	Crustacea: mechanical strength
	Plants (leaves): calcium storage
Calcium hydroxyapatite $[Ca_{10}(PO_4)_6(OH)_2]$	Vertebrates (bones): mechanical strength and ion store
	Mammals (teeth): cutting, grinding
	Fish (scales): protection
Amorphous Octacalcium phosphates $[Ca_8H_2(PO_4)_6]$	Vertebrates: bones and teeth
	Chittons: teeth
	Bivalves, Mammalia, Cow: Ca ion storage
Gypsum ($CaSO_4.2H_2O$)	Jellyfish: gravity receptor
Weddellite ($CaC_2O_4.nH_2O$)	Plants: ion store
Calcium pyrophosphate ($Ca_2P_2O_7$)	Gastropoda: detoxification
Magnesium calcite $(Mg,Ca)CO_3$	Octocorralia (spicules): mechanical support
	Echinoderms (spines): strength and protection
Amorphous silica ($SiO_2.nH_2O$)	Diatoms, Radiolaria: skeleton
	Plants (leaves): protection
Magnetite (Fe_3O_4)	Bacteria/Salmon: magnetic navigation
Ferrihydrite $[Fe_2O_3.nH_2O]$	Plants, (in)vertebrates (ferritin): storage protein
Greigite (Fe_3S_4)	Bacteria: magnetic navigation

material, while limy coats are translucent. Therefore, a silica cover is more useful to support photosynthesis in diatoms. At this point, it is worth noting that most biominerals are ionic in nature; however, silica-based minerals feature Si-O-Si linkages, which in the presence of water give rise to hydrated inorganic polymers that are molded into elaborate shapes by many unicellular organisms [6]. Furthermore, the widespread appearance of siliceous biominerals as structural support in various plants and animals suggests the essential nature of silicon.

Among transition elements, Fe and Mn participate in the biomineralization process. The chemistry of biominerals consisting of these elements is dominated by the redox behavior between the 2+ and 3+ oxidation states. Iron oxide biominerals are employed to strengthen soft tissues and as storage reservoirs. Besides this, the magnetic properties of magnetite are utilized by various magnetotactic bacteria as a means of navigation.

Other than the above-mentioned elements, the sulfides of Cu, Zn, and Pb are also found on the external surfaces of bacteria, while some yeasts mineralize intracellular CdS nanoparticles [7]. Localized deposition of Alaskan gold nuggets on the surface of bacteria is the only known biomineralization process that involves the formation of Au crystallites.

It has become evident that the majority of biominerals are formed using abundant elements in the Earth's crust under mild conditions, at near-neutral pH and ambient temperature [6]. However, it is crucial to gain a deeper understanding of the mechanisms behind biomineralization and the role organic molecules play in the formation of inorganic materials in biological environments.

AMORPHOUS AND CRYSTALLINE BIOMINERALS

Biomineralization is the process that leads to the formation of hierarchically structured organic-inorganic composites mediated or influenced by living organisms. Biominerals are extremely prevalent in nature. To date, over 60 biominerals have been identified among the five animal kingdoms. These biominerals can be broadly classified into three categories based on the degree of crystallinity:

i. *Amorphous biominerals:* These materials are also called frozen liquids and lack preferred growth direction and morphology. Due to this, the amorphous structure readily settles into the form of any mold. Moreover, they do not have planes and are significantly less brittle than crystalline compounds. Amorphous solids are commonly observed in oxide elements with intermediate electronegativity [8]. Due to the intermediate electronegativity of elements, the bonding can be considered partially ionic and partially covalent. Common examples of this category include oxides of Groups III, IV, and V. This nature of bonding facilitates these oxides to form three-dimensional polymeric scaffolds joined through corners instead of edges or faces. The polymeric nature of these oxides is responsible for the low solubility and high viscosity of these materials. Biomineralization of amorphous silica by diatoms is a classic example of this class. Other examples include calcium carbonate, calcium phosphate, iron hydroxides, etc., which are widespread in nature. One interesting thing regarding amorphous calcium carbonates and phosphates is their tendency to crystallize into a number of polymorphic forms, which is favored both thermodynamically and kinetically. Despite this, the presence of the amorphous form represents a deliberate effort by organisms to inhibit crystallization [8]. Amorphous biominerals are commonly used as precursors for crystalline forms, temporary storage sites for metal ions and repositories for embedding toxic metals, which are then eliminated together from the cell. As we progress further in this chapter, we will delve into a deeper understanding of the role of amorphous materials in biological systems.

ii. *Polycrystalline biominerals:* Polycrystalline biominerals are characterized by the presence of multiple crystallites of varying sizes and orientations, separated by grain boundaries. The formation of crystalline materials typically occurs through the creation of high local supersaturation, followed by nucleation. As a result, rapid crystallization often gives rise to polycrystalline structures. These crystallites can be effectively organized to form hierarchical structures and crystalline arrays. Moreover, this class of biominerals exhibits a wide range of morphologies. An example of a polycrystalline biomineral is mollusk nacre, which is composed of polycrystalline calcium carbonate and is renowned for its remarkable mechanical strength.

iii. *Single crystal biominerals:* Single crystals are materials in which the crystal lattice of the entire sample is continuous with no grain boundaries. These materials possess a unit cell that represents a fundamental spatial and symmetry relation between atoms and molecules. The formation of the entire structure from a single crystal offers some unique advantages, such as maximal packing, excellent order, and homogeneity. The external form of the single crystal is a reflection of its internal symmetry. Generally, single crystals have sharp edges and flat sides. However, biomineral-derived single crystals often display remarkably unconventional forms that have evolved to enhance their specific functions [9]. For example, the teeth of sea urchins deviate from the typical flat sides and sharp edges observed in conventional single crystals. Instead, sea urchin teeth exhibit an extraordinary arrangement of intertwined curved plates and fibers that interlock and fill the teeth as they grow. This intricate structure, composed of calcite, imparts exceptional hardness, enabling the sea urchin to grind rocks. Remarkably, this complex architecture arises from a single nucleation event.

Another remarkable aspect that distinguishes biomineral single crystals from their inorganic counterparts (crystals produced in vitro) is the presence of occluded macromolecules within the crystal lattice. These embedded macromolecules have a significant effect on the shape, morphology, and, most importantly, the mechanical properties of biomineral single crystals. For instance, single crystals of the $CaCO_3$ calcite phase can be easily cleaved along the (104) plane. In contrast, sea urchin skeletal elements also formed from calcite (biomineralized), fracture with difficulty. Inspection of the same revealed that acidic proteins occluded inside the mineral phase are responsible for this fracture tolerance in the mineralized form. Acidic proteins introduce this fracture tolerance in sea urchin skeletal elements and other biominerals by controlled intercalation on planes oblique to the cleavage planes [8].

It is estimated that, out of all the biominerals identified, 20% are amorphous and the remaining 80% are crystalline in nature. It is also possible that amorphous biominerals are underestimated because of the difficulty in characterizing them.

MECHANISM OF BIOMINERALIZATION

There are four processes in biomineralization or crystallization that need to be carefully controlled to obtain the desired products. These are the solubility of constituent ions, level of supersaturation, nucleation and crystal growth. In the case of biomineralization, the biominerals are in constant interaction with biological molecules such as proteins and phospholipids present in the biological membrane [9]. Biomineralization is initiated by the precipitation of the constituent ions due to their supersaturation in the solution. Biomineralization falls into the category of heterogeneous nucleation, where nucleation occurs at specific sites (e.g., proteins or membranes, etc.) provided by the microorganisms. These specific sites at the surfaces or within the cells lower the surface energy, which eventually decreases the activation energy required for crystal growth. The most accepted model for biomineralization is the LaMer model from the 1950s, proposing the concept of burst nucleation, initially conceptualized to explain nanoparticle synthesis. The mechanism is displayed in Figure 10.2 and is as follows: (i) The process is initiated with the concentration of constituent ions and continues until their concentration reaches a certain critical supersaturation level (C_s) where nucleation is possible. (ii) The saturation level continues to increase until it reaches a level (C_{min}) at which the activation energy barrier can be easily overcome, leading to the burst of nucleation. (iii) Because of the burst of nucleation, the supersaturation level decreases below the auto-nucleation level, marking the end of the nucleation process and the initiation of the growth process [10]. More ions/particles migrate toward the nucleated particle surface, leading to mineralization. According to this model, the concentration of constituent ions/particles over time increases rapidly during the self-nucleation stage (ii) and then slows down or remains constant during the final growth stage (iii).

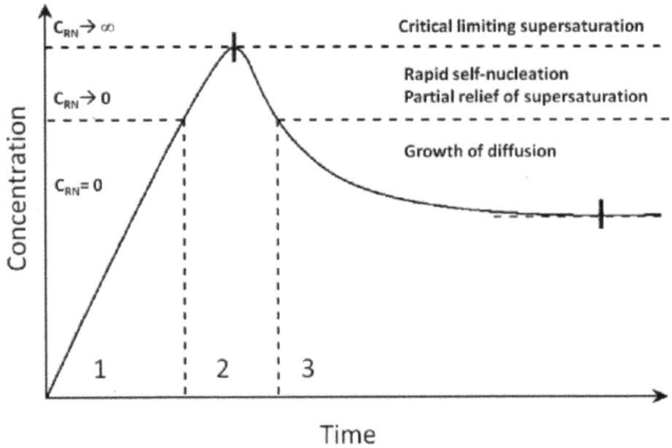

FIGURE 10.2 Schematic representation of LaMer model describing nucleation and nucleation growth. (Copyright permission from American Chemical Society, Ref. [10].)

TYPES OF BIOMINERALIZATION

Biomineral formation initiates with the nucleation step, followed by crystal growth. All this takes place within a well-defined, spatially delineated site [7]. In the case of unicellular organisms, these sites can be epicellular (i.e., on or within the cell wall) or intracellular. Multicellular organisms have evolved specialized cells to regulate the biomineralization process in extracellular compartments [7]. These isolated regions must regulate diffusion into or out of the system. To maintain the electrical neutrality of the fluid within the region, ion supply occurs by two modes: either through active pumping or by a passive diffusion gradient. Depending on the degree of biological control and the mechanism of nucleation and growth (Figure 10.3), biomineralization can be broadly classified into two fundamentally different classes: (i) Biologically Induced Biomineralization (ii) Biologically Controlled Biomineralization.

BIOLOGICALLY INDUCED BIOMINERALIZATION (BIM)

In this class of biomineralization, mineralization occurs as a consequence of the interaction between biological activity and the environment. The microorganism cell is a causative agent only and exerts very little control over the mineralization process, i.e., it inertly contributes to mineral formation. However, the pH, pCO_2, and composition of secretion products are regulated by the organisms as a result of their metabolic activity within a particular redox environment. Additionally, they also facilitate mineralization by providing a charged cell surface or reactive site. By providing reactive surfaces, microorganisms induce precipitation in supersaturated solutions. The surfaces are rich in ionized ligands that can promote chemical reactions like complexation, reduction, and precipitation, which assist in BIM. The negatively charged surface of microorganisms concentrates the cations

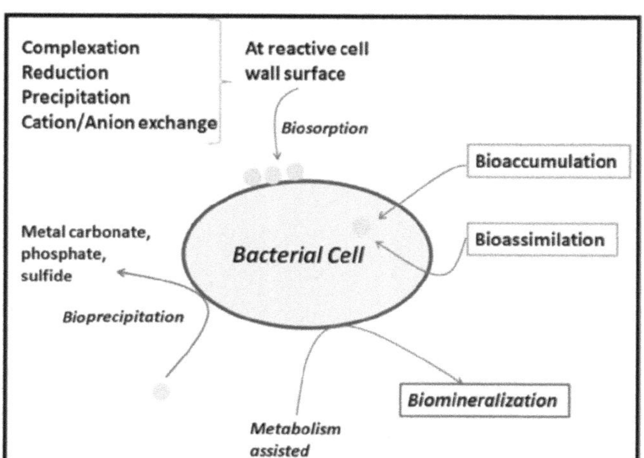

FIGURE 10.3 Various modes of biomineralization.

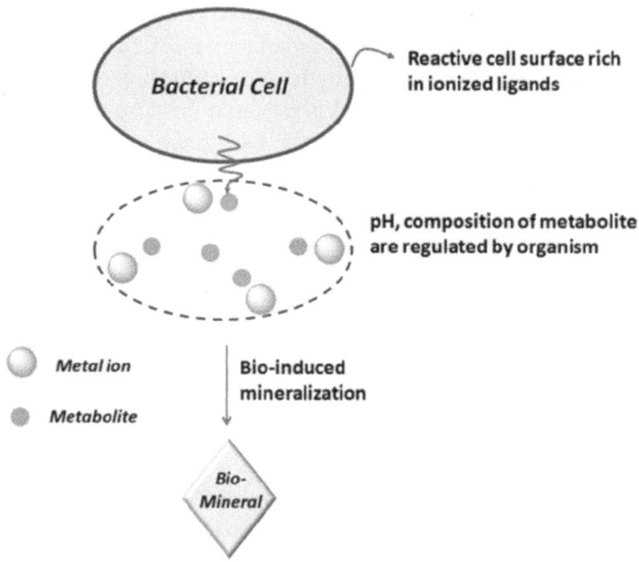

FIGURE 10.4 Illustration for metabolite-assisted biomineralization under BIM.

over the surface, where they are subsequently coordinated by the anions from the external environment, thereby initiating the mineralization process. Another pathway for BIM involves the metabolic by-products of microorganisms. The by-products facilitate mineral growth by providing suitable ligands that can react with ions in the external environment, which might not be possible without the presence of these microorganisms. Heterogeneity in terms of variable external morphology, structure, particle size, water content, and trace element composition is a characteristic phenomenon of this class of biomineralization. The composition of biominerals obtained through BIM shows large variations depending upon the chemical nature of the environment within which they form [11]. Figure 10.4 provides a schematic representation of bio-induced mineralization. Prominent examples of BIM are sulfide minerals that require sulfate-reducing bacteria for their growth. Sulfate-reducing bacteria absorb, concentrate, and reduce sulfate to sulfide, which is released as a by-product. This sulfide reacts with the ions present in the external environment leading to the formation of sulfide minerals.

Biologically Controlled Biomineralization (BCM)

It is a process where microorganisms actively control the nucleation, growth steps, morphology, and location of the biomineral formed to serve a physiological purpose. Since this class of biomineralization is actively controlled by organisms, the minerals can form even when the environmental conditions are not thermodynamically favorable for mineral formation. This confirms that microorganisms not only catalyze or initiate mineral formation but also control the process by creating a microenvironment thermodynamically favorable for a particular mineral to precipitate. In other words, microorganisms in BCM exert control at each and every step of mineralization, leading to very controlled minerals in terms of structure and morphology. This control is essential because microorganisms use these minerals for specific purposes, thus their structure and morphology have to be very precise.

In BCM, microorganisms use either intracellular or extracellular organic matrices to bind particular ions for mineralization. The extracellular and intracellular classification is with reference to the cell responsible for mineralization. However, classifying BCM is not that simple. Sometimes, biomineralization is initiated within the cell and then proceeds in the extracellular space. Nevertheless, for fundamental understanding, we will discuss extracellular and intracellular BCM separately.

In intracellular mineralization, microorganisms isolate the locus of mineralization and start accumulating ions within specialized vesicles or vacuoles. Afterwards, the accumulated metal ions are exposed to organic ligands in a controlled way, which initiates the mineralization process, as shown in Figure 10.5. In such cases, the cell exerts a high degree of control over the concentration of constituent ions of biominerals as well as over the environment and concentration of trace elements within the specialized compartment of biomineralization. The biomineral crystallite thus formed may leave the cell as an individual unit, or it may get pre-assembled before secretion. Examples belonging to this class include calcite biomineralization to form the coccolith structure. The released biomineral can be immediately employed, as in the case of coccoliths, or can mark the beginning of a secondary growth process wherein the secreted biomineral interacts with an extracellular complex organic matrix to form an ordered extracellular structure with preferred crystallographic orientation, as in the case of spicules of sea urchins [12]. Although rare, one more effective strategy is observed in intracellular BCM, wherein the intracellular product does not become extracellular.

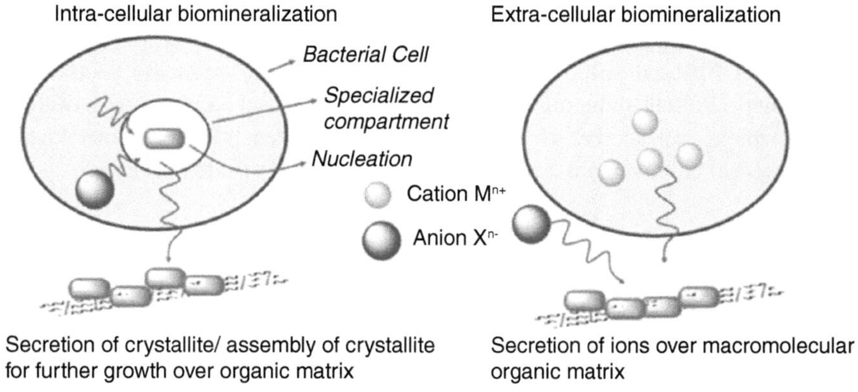

FIGURE 10.5 Schematic representation of intracellular and extracellular biologically controlled mineralization processes.

An example of this class is the formation of magnetosomes in magnetotactic bacteria. The magnetite biomineral formed remains membrane-bound and assembles through its magnetic fields to produce the biomineral chain, as shown in Figure 10.1f.

In extracellular BCM, the cell transfers the constituent ions of the biomineral to the organic matrix, which becomes the site of nucleation and crystal growth, as shown in Figure 10.5. The transport process can occur through the active pumping of ions by the cell or the concentration of ions in vesicles, which will eventually be secreted into the extracellular space. The organic matrix, which acts as a site of biomineralization, is a three-dimensional structure that consists of biomolecules like proteins, polysaccharides, etc., which perform vital functions involved in the formation of hybrid or composite biominerals. The composition of this organic matrix is genetically programmed and specific to each species [1]. Aragonite formation in mollusk shell nacre belongs to this class of mineralization.

Apart from extracellular and intracellular BCM, the third type of mineralization process is 'intercellular BCM'. In this process, a community of single-celled organisms exploits sites between their cells rather than within the cells for mineralization. The cell surface controls the nucleation and growth of the biomineral and directs the polymorph and shape of the biomineral that forms. The formation of calcite sheets by calcareous algae with c-axis orientation perpendicular to the cell surface is an example of this class of mineralization.

It is important to understand that, unlike BIM, where mineralization occurs at the surface of microorganisms, in BCM, place within the organism. However, there is one similarity between the BIM and BCM processes: organisms often produce metastable amorphous rather than crystalline minerals under low-temperature conditions, unlike abiotic mineralization, which takes place at higher temperatures and thus produces crystalline material.

ROLE OF ORGANIC MOLECULES IN BIOMINERALIZATION

It is increasingly recognized that organic materials play a crucial role in the construction of biominerals that perform various biological functions, including structural support, protection, and signaling. Organisms do not alter temperature or pressure to produce materials of specific size, morphology, structure, and orientation, as commonly practiced in synthetic chemistry. Rather, they exert control over the mineralization process through organic molecules that include proteins, lipids, nucleic acids, and carbohydrates. These organic molecules construct an organic matrix framework with nano-confined spaces that act as molds for the selective precipitation of minerals with defined morphology and structure, control ion input, provide nucleation sites, and exert fine control over crystal orientation and growth. Lastly, they also act as inhibitors by terminating the crystal growth of a particular polymorph [1]. The celebrated biologist, mathematician, and scholar Sir D'Arcy Thompson, in his book *On Growth and Form*, provides a scientific explanation behind the patterns and body structures formed in plants and animals [9]. According to him, there is a direct correlation between biological forms and

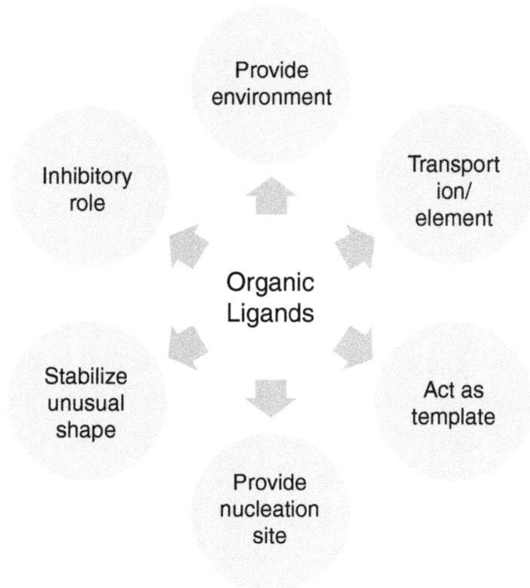

FIGURE 10.6 Various crucial roles performed by organic ligands during biomineralization process.

mechanical phenomena. Biominerals adopt their shape due to the assembly of cells, which depends upon the packing and relative surface tension between the cells. The various functions performed by organic ligands during biologically controlled crystal growth are summarized in Figure 10.6.

Role of Organic Matrix

Biologically controlled biomineralization occurs within a structured organic matrix having designated spaces for crystal growth. This organic matrix can consist of lipid vesicle, as in the mineralization of magnetite (Fe_3O_4) by magnetotactic bacteria, or collagen fibril macromolecular frameworks for the growth of hydroxyapatite, both of which are vital for organisms to exert control over crystal growth. These organic molds direct the sites for nucleation and allow organisms to regulate the composition of the solution in terms of desired minerals and additional additives. Furthermore, the spatial constraints imposed by the organic molds are key factors in determining the final morphology of the mineral [1]. In some cases, the organic molds are further functionalized by the adsorption of soluble macromolecules that can alter the rate of nucleation. These macromolecules may also act as soluble additives, which can adsorb onto the crystallites during growth, thereby modulating the morphology and texture of the mineral [13]. The presence of macromolecules within the mineral, up to a few weight percent, has been observed in several cases, including calcium carbonate polymorphs, hydroxyapatite, silica diatoms, and sponge spicules. Isolation of these intra-mineral macromolecules in the case of hydroxyapatite revealed a sequence of polysaccharide-aspartic acid with intervening glycine. This sequence bears a net negative charge, which is considered responsible for binding of Ca^{2+} ions and plays a significant role in controlling crystal growth [14]. Similarly, the chemical groups present on the silaffin proteins (long-chain polyamines) and silicatein proteins (hydroxyl-rich structures) extracted from diatoms and silica sponges are postulated to be important in the silicification process (precipitation of silica).

Role of Ion Pumps and Channels

Another interesting point to note is that organisms precipitate minerals from environments where these minerals are typically undersaturated, yet the organisms are still able to mineralize them. This ability is attributed to the presence of various ion-specific pumps and channels on the cell membrane. These pathways help determine the composition as well as the concentration of ions at the site of action (i.e., the mineralization site). Additionally, they can dictate the order of introduction of ions at the site. Many minerals, such as iron oxides and calcium carbonates, exist in multiple phases or polymorphs. Organisms exert fine control over the selective synthesis of a particular phase by regulating the concentration of ions, the pH of the solution, and through interactions with surrounding macromolecules. For example, in the case of magnetotactic bacteria, the magnetite minerals responsible for the geomagnetic navigation of bacteria in their aquatic habitat crystallize within specialized intracellular membranous compartments called magnetosomes, synthesized by the bacteria. The diameter of magnetosomes typically varies between 20 and 100 nm. Readers are encouraged to refer to Ref. [14] to understand the various stages involved in the controlled synthesis of magnetite by magnetotactic bacteria. It has been revealed that several proteins are involved in the formation and maintenance of membranous vesicles. One such class of proteins is the Mms proteins, which likely play vital roles in the accumulation of iron, nucleation of iron oxide, and regulation of redox behavior and pH to exert control over the crystal size and shape of magnetites [14,15].

The crucial role of Mms proteins has been elucidated through controlled experiments. The synthesis of magnetite in the absence of Mms proteins resulted in smaller and irregularly shaped nanocrystals, as shown in Figure 10.7. The Mms proteins exhibit a common amphiphilic character, having hydrophobic N-terminal and hydrophilic C-terminal regions. Interestingly, these proteins are exclusively found in magnetotactic bacteria. It is postulated that the hydrophilic C-terminal region of the protein is an iron-binding site, while the hydrophobic N-terminal region is responsible for the self-aggregation of the protein [16].

Role of Nanoscale Pores and Proteins in Regulating Orientation of Biomineral

The vital role of organic molecules in the biomineralization process can also be appreciated from the crystal growth of hydroxyapatite, also called bioapatite, a biologically produced analog of hydroxyapatite in bones and teeth. In bone formation,

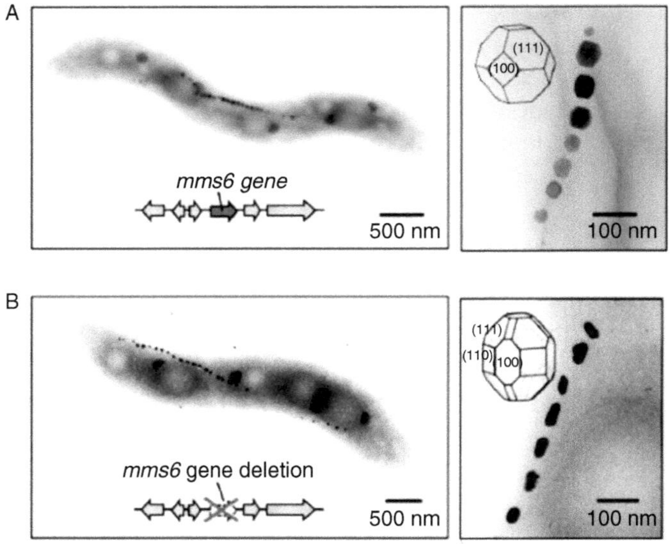

FIGURE 10.7 Effect of the presence and absence of Mms protein on the size and shape of magnetite mineralized in magnetotactic bacteria. (This figure is under Creative Common Attribution and is adopted from Ref [14].)

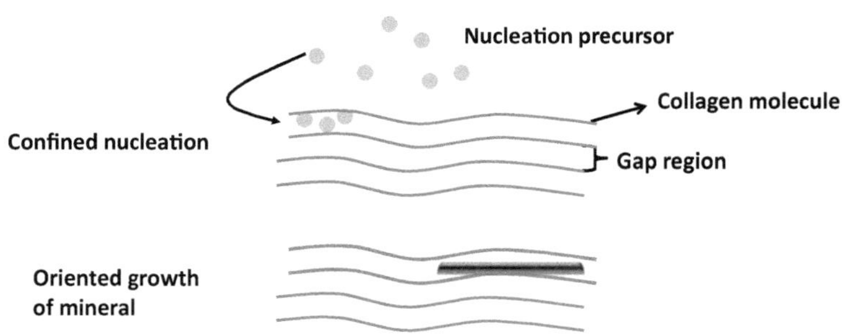

FIGURE 10.8 Schematic representation illustrating the role of nano-confined gaps in collagen fibers in oriented growth of mineral.

collagen fibrers act as templates for the deposition of hydroxyapatite crystals. The collagen fibres consist of staggered packing of individual collagen molecules in a way that periodic nanoscale pores or gaps are created on the fibril surface and within the channels [17]. These nano-confined gap regions within the collagen fibrils decrease the energy barrier for the nucleation of hydroxyapatite. The nano-gap regions of collagen have a net positive charge, which electrostatically attracts net negatively charged amorphous calcium phosphate nuclei and initiates the nucleation process. Also, this nanoconfined gap geometry guides the morphology and structure of hydroxyapatite crystals, as schematically depicted in Figure 10.8. In addition to collagen fibrils, phosphoproteins also play a vital role in mineralization. Subsequent to the formation of the collagen matrix, phosphoproteins are secreted within the matrix. Studies have shown that for the mineralization of hydroxyapatite crystals, interaction between phosphoproteins and collagen is essential to induce mineralization [18]. Besides this, phosphoproteins also control the rate of crystallization of calcium phosphate within the bone. It has also been found that mineralization of hydroxyapatite crystals in the unconfined regions (extrafibrillar spaces of collagen) results in aggregates without specific orientations [17].

Role of Organic Macromolecules in Polymorph Selection

One of the major challenges in the field of biomineralization is to understand and unravel the mechanism(s) by which biological systems select, which polymorph will mineralize. Selective precipitation of polymorphs occurs almost with 100% fidelity. The role of organic ligands plays a very crucial role in determining the precipitation of a particular polymorph. This can be understood from the biomineralization of calcium carbonate, which is one of the most abundant inorganic biominerals. Along with hydroxyapatite, calcium carbonate constitutes the basic building blocks available for skeletal structures. Nature has employed these building blocks to produce diverse structures with unbelievable morphology, as presented elsewhere [3,19]. The diversity in the bio-architecture is attributed to the spatial constraints imposed by the matrix and the role of organic ligands involved in the biomineralization of calcium carbonate [3].

Calcium carbonate is known to exhibit six polymorphic forms, including five crystalline and one amorphous form. Out of the five crystalline forms, three are pure calcium carbonate, while two are hydrated forms of crystalline calcium carbonate. Pure crystalline calcium carbonate exists in three polymorphic forms: calcite, aragonite, and vaterite. These polymorphs have different crystal structures, which influence their physical and chemical properties. For instance, calcite crystallizes in trigonal symmetry with space group R-3c. Different from calcite, aragonite adopts an orthorhombic crystal structure with Pmcn space group. Ca^{2+} in calcite is octahedrally coordinated by oxygen atoms, whereas in aragonite, the coordination around Ca^{2+} is nine-fold, defined by carbonate oxygen, as shown in Figure 10.9 [20]. On the other hand, the vaterite form of $CaCO_3$ stabilizes in the hexagonal crystal system and belongs to the $P6_3/mmc$ space group. Ca^{2+} ions form a primitive hexagonal array wherein they reside at the corners of trigonal prismatic interstices, octahedrally surrounded by oxygen from the carbonate ion.

Calcite is the most stable polymorph of calcium carbonate, followed by aragonite. However, they have little difference in their relative stabilities. On the other hand, the vaterite form of calcium carbonate is metastable and is considered to be the intermediate between amorphous calcium carbonate (ACC) and stable crystalline polymorphs. These polymorphs have been mineralized by organisms mainly to form hard structures, such as shells and exoskeletons. Shells of mollusks and sea snails are mainly composed of calcite or aragonite. Besides this, polymorphs are also used to perform some specific functions. Aragonite is used as a light sensor in some microorganisms. Some insects use calcite in their eyes because of its unique optical properties, like birefringence and double refraction,

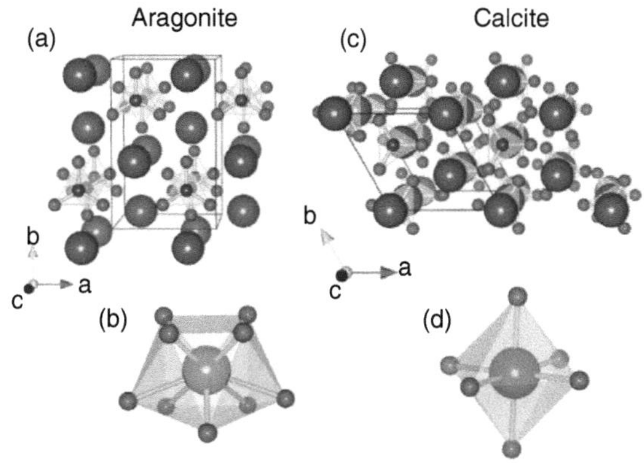

FIGURE 10.9 Schematic illustration of crystal structures of calcite and aragonite. (This figure is under Creative Common Attribution and is adopted from Springer Nature, Ref. [20].)

to detect polarized light patterns in the sky, which help them navigate and orient themselves. Though not as commonly found as aragonite and calcite, vaterite also plays an important role in the biological processes of some organisms, like vaterite crystals found in the stomachs of some birds, where they act as a pH buffer and neutralize acidic content. Besides these crystalline phases, a large number of microorganisms like echinoderms, mollusks, etc., employ ACC along with crystalline forms. ACC acts as a pH buffer, is involved in the repair of damaged or broken mineralized tissue, and most importantly, is thought to act as a precursor or template for crystalline phases and regulate the mineralization of crystalline carbonate structures like exoskeletons and shells.

Experiments under ambient conditions have revealed that most of the time it is the calcite phase that crystallizes from the solution compared to aragonite. The abundance of aragonite biomineral, despite being less stable than calcite, is puzzling for most scientists. Nature has demonstrated a unique ability to selectively mineralize stable or metastable phases of calcium carbonate. For this, nature first isolates the environment of mineral formation from the outside world and then employs biological ligands (organic molecules) during crystal growth, which can inhibit the nucleation of a particular form or prefer the nucleation of a metastable form [21]. The organic molecules can also dictate the selective crystallization of calcite or aragonite, which have only a slight difference in their relative stability, as clearly evident from mollusks. For calcium carbonate shell formation, these microorganisms construct a matrix comprising various macromolecules such as polysaccharide β-chitin, hydrophobic silk protein, and a complex collection of hydrophilic proteins. During mineral formation, some of these proteins get trapped in the crystal [2].

To understand the role of proteins in polymorph selection, in-vitro experiments have been carried out involving the crystallization of calcium carbonate within the highly structured layers of β-chitin and hydrophobic silk fibroin in the presence of soluble hydrophilic protein organic macromolecules, which are the main component of nacreous layers of mollusks. The soluble macromolecules used in the experiments were extracted either from the calcitic mollusk or the aragonite mollusk. The outcome revealed that product formation is highly dependent upon the nature of soluble macromolecules. Calcitic macromolecules induced calcite growth inside the chitin layers, whereas aragonitic macromolecules facilitated aragonite crystallization [22]. Close inspection of the chemical nature of these macromolecules revealed that these are acidic in nature and particularly rich in aspartic acid. Further, the mole percent of aspartic acid is higher in calcitic macromolecules compared to aragonitic macromolecules. In other words, the macromolecules involved in the formation of calcite mineralization are strongly polyanionic and more strongly acidic than macromolecules responsible for aragonite formation. The strong polyanionic nature may have provided a strong binding site for Ca^{2+} ions, thereby facilitating local super-saturation.

A clear illustration of the vital role of soluble proteins in directing crystal structure formation comes from studies pertaining to the precipitation of amorphous calcium carbonate. Analysis of ACC in sponges and ascidians showed the presence of some glycoproteins within it. These glycoproteins are found to be rich in glutamic acid and/or glutamine. Formation of calcium carbonate in the presence of these glycoproteins resulted in ACC. Subsequent studies showed that ACC in some cases acts as a precursor phase of both calcite and aragonite. More interestingly, organisms have also evolved strategies to control the transformation of ACC into calcite or aragonite phases.

The organic matrix comprising polysaccharide β-chitin layers, hydrophobic silk fibroin protein, and associated specific acidic proteins not only facilitates the selective phase formation of crystalline calcium carbonate but also promotes oriented crystal growth, which is a key feature of many biominerals. To understand this aspect in further detail, we shall look at the aragonite crystals of the mollusk shell as shown in Figure 10.10a.

Experimentally obtained information through detailed microscopic and diffraction measurements showed that hydrophobic silk protein in the form of hydrated gel is located between polysaccharide chitin layers, and acidic proteins are located at the surface of the chitin layer, acting as a site for nucleation of aragonite crystals. It has also been observed that there is a strong correlation between the chitin layer and crystallized aragonite mineral, which indicates that the nucleation mechanism is possibly epitaxial in nature (oriented crystal growth of new crystalline layers with respect to the substrate matrix). The binding of Ca^{2+} ions onto the surface of oriented proteins within the organized layered matrix induces sheet-like orientation during crystal growth. It can be considered a brick-and-mortar-like structure in which aragonite crystal sheets are held together by soluble proteins [23]. In in-vitro experiments, it is observed that aragonitic proteins induced aragonite crystal formation with needle-like morphology. In the presence of a mixture of calcitic and aragonitic proteins, aragonite crystal formation takes place in the form of polycrystalline plates, having oriented stacks of crystals. Thus, from the above discussion, it is quite apparent that soluble

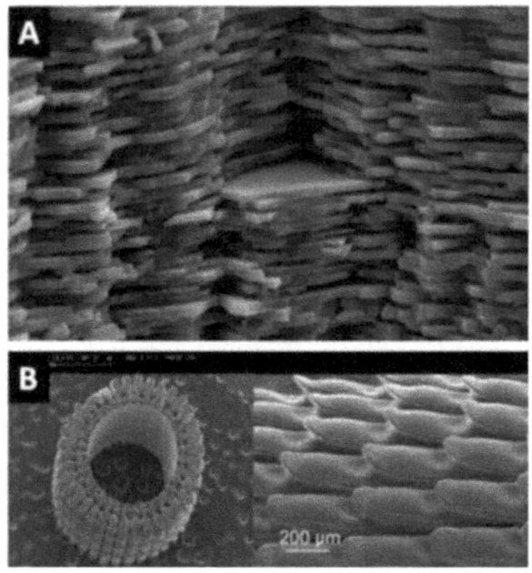

FIGURE 10.10 Images depicting the morphology of (a) Highly oriented sheet-like aragonite nacre, (b) Structure of long hollow sea urchin spines arising from the assemblage of skeletal plates comprising single crystals of calcite. (Copyright (a) under creative common license; credit Fabien Heinemann and (b) This figure is under Creative Common Attribution and is adopted from PLOS, Ref. [23].)

macromolecular protein plays a key role in deciding the phase and morphology of biominerals [8,23]. These soluble macromolecules are believed to be adsorbed on the specific crystal faces, thereby producing modified crystal morphology. The specific adsorption arises due to molecular-level recognition between the chemical motifs of macromolecules and crystal faces, which stabilize these faces. This interaction subsequently leads to the formation of a hybrid inorganic-organic system wherein these soluble macromolecules get embedded into the mineral, causing significant modification to the crystal's morphology. The evidence of the interaction of these macromolecules with selective faces of growing crystals has been obtained from in-vitro experiments dealing with atomic force microscopy (AFM) imaging of rhombohedral calcite crystals in saturated calcium carbonate solution. The images depict the presence of sharp straight edges and corners of the crystals, which become round upon exposure to intracrystalline proteins extracted from the calcitic layer. However, the action of soluble proteins alone is not sufficient for producing crystals with such unique morphology, as shown in Figures 10.1 and 10.10.

Organic Macromolecules Governing the Morphology of Biominerals

Organisms have evolved strategies to produce crystals wherein the overall morphology is quite different from the symmetry of the crystal lattice. Closer inspection of sea urchin spines reveals a sponge-like biocontinuous structure arising from the assemblage of skeletal plates. X-ray diffraction and other studies reflect that these skeletal plates are composed of single crystals of calcite. These single crystals grow in a biocontinuous manner with curved surfaces. A number of factors, such as the activity of ion pumps and channels and the storage granules of ions, are also responsible, along with the interaction of soluble macromolecules embedded within many biominerals, for this unique morphology. Furthermore, given that crystals with such unique morphologies are usually formed within vesicles with defined shapes that may be altered during the crystal growth process, the role of spatial constraints imposed by the vesicles cannot be ignored. In some cases, it is also observed that ACC plays a crucial role in the crystal growth of calcium carbonate. It acts as a precursor phase and exerts control over the morphology of the crystal. Initially, tiny rhombohedral calcite crystals are formed, which continue to grow by the addition of ACC. The impinged ACC on calcite crystals gradually transforms into single crystals; however, the morphology is imposed by the environment and not the crystallographic morphology. In-vitro experiments also reveal that the precipitation of calcite at room temperature in the absence of ACC results in the isolation of calcite crystals with irregular morphologies, whereas in the presence of ACC, they acquire shape imposed by the matrix [24].

Another interesting phenomenon is the assemblage of single-crystal components generating polycrystalline structures with complex morphologies, as observed in the coccolith scales of coccolithophores. Despite their small dimensions of a few micrometers, coccoliths are remarkably elaborate biominerals that display fine control of the organic system over both the nucleation and growth of calcite. The coccolith scales consist of 30–40 units of calcite crystals oriented in a clockwise direction to produce a double-rimmed structure with morphological chirality. The formation of coccoliths represents an example of intravesicular biomineralization, wherein nucleation occurs within vesicles derived from the Golgi body. Prior to nucleation, the formation of an organic base plate scale takes place with a characteristic microfibrillar structure. Nucleation of calcite occurs at the periphery of this base plate to form the protococcolith ring of calcite crystals, followed by oriented growth upward and outward to form the overall coccolith structure [25]. One unique observation during crystal growth is that the shape of the vesicle constantly changes to maintain close contact with the crystals, which is supposed to be crucial in producing the distinctive pattern.

Organic Macromolecules and Mechanical Strength of Biomineral

Commonly observed in biominerals is their impressive mechanical strength, which arises due to their organic-inorganic hybrid nature and structural organization. For instance, when mollusk nacre is compared with artificial ceramics or crystals, nacre exhibits superior performance in terms of strength, stress resistance, and fracture tolerance [26]. These excellent mechanical properties of mollusk nacre are attributed to its structural features. As already described, mollusk nacre comprises aragonite crystal platelets sandwiched between organic materials, forming a hierarchical structure. This structure, reinforced by the organic matrix, is believed to be the key to the fracture toughness of nacre, which is 3,000 times that of bulk aragonite. The organic layer surrounding the aragonite platelets helps in energy dissipation and imparts mechanical anisotropy to the nacre, which prohibits the transmission of perpendicular cracks. Additionally, the surface of aragonite platelets features nano-extensions called asperites, which act as mineral bridges between two platelets and impart sheet resistance properties to nacre. Gaining inspiration from the remarkable properties exhibited by natural composite materials, researchers have been making efforts to mimic these materials for various applications [27].

BIO-INSPIRED SYNTHESIS

As we have seen, living organisms can intricately control the deposition of inorganic minerals in association with organic matrices that serve a variety of purposes, such as navigation, optics, skeletal formation, and protection. These biominerals have elaborate hierarchical composite structures with unique morphologies and incredible physical and chemical properties that surpass those of their synthetic counterparts. These intricate structures have captivated chemists, physicists, and materials scientists in their quest to develop novel materials with exceptional properties and functionalities. Furthermore, this has also inspired material chemists to rethink conventional material synthesis strategies. One of the most important aspects of biomineralization is the synthesis of complex structures in aqueous media under ambient conditions. By emulating the principles of biomineralization, material chemists can create materials with enhanced mechanical strength, stimuli-responsive characteristics, self-healing abilities and biocompatibility while reducing ecological footprints compared to conventional synthesis [14,28].

Organisms employ several strategies, such as the occlusion of proteins inside crystals, the organization of crystals into large superstructures, embedding inorganic crystals into organic matrices, arranging the mineralized organic matrix to generate diverse structures, and more. The basic objective of organisms is to exert spatial, chemical, structural, morphological, and constructional control to generate highly sophisticated minerals adapted to specific functions [28].

To elaborate further, the spatial control of organisms refers to creating delineated enclosed spaces specifically designated for biomineralization. Within these spaces, ions and other components of biominerals are selectively transported, which imparts control over the growth kinetics of the crystals [1].

Regulating physicochemical factors such as the solubility of precursors, ion transport, nucleation, and crystal growth comes under the purview of chemical control. Organisms regulate these factors by controlling the movement of ions in the space delineated for biomineralization and through macromolecules like polysaccharides and glycoproteins, which often function as facilitators or inhibitors of nucleation and crystal growth [22].

Quite often, an organic matrix functionalized with chemical entities functions as a template during biomineralization, thereby controlling nucleation as well as the growth of particular crystal faces or axes. This refers to the structural control organisms exert over the biomineralization process [28].

Through morphological control, organisms dictate the specific morphology of biominerals. This is achieved by growing minerals in enclosed spaces where proteins interact with specific crystal faces based on molecular recognition, thereby altering growth kinetics. Spatial confinement ensures that crystal growth is controlled by the organism rather than the intrinsic crystallographic axes of the unit cell. Additional control is exerted by providing a hydrogel-like environment inside specialized compartments as seen in the formation of aragonite within the chitin layer, where silk fibroin protein creates a hydrogel-like environment [2]. In several cases, hierarchical structures comprising assemblies of crystals have been observed. The formation of these structures is possible due to the constructional control exerted by organisms.

Many *in vitro* experiments have been carried out to understand and mimic the role of various macromolecules employed by organisms during mineral nucleation and crystal growth. These include growing crystals over surfaces with acidic functionalities and deciphering the role of interactions and orientations of these acidic groups with growing minerals in regulating nucleation and preferred crystal growth. The outcome of these experiments revealed that the adaptation of functional groups present on surfaces to align with the orientation of growing crystals is a vital factor in regulating nucleation [29]. In another experiment, mineralization in the presence of proteins extracted from biominerals such as magnetosomes, mollusk shells, bone apatite, and others has been investigated. These proteins exhibit excellent control over crystal growth and morphology in *in vitro* experiments [16]. The function of these proteins has been successfully emulated by synthetic polymers such as polyaspartic acid and polyacrylic acid to regulate magnetite formation and control the polymorph and morphology of calcium carbonate crystals [28]. Most interestingly, polymer track-etch membranes can also impose spatial control on growing calcium carbonate crystals to mimic the morphology of biominerals. Experiments dealing with the effect of a gel-like environment during crystal growth have shown that such environments not only change crystal morphology by altering kinetics but also lead to gel inclusion within the crystals, which improves shock tolerance [30].

There is now sufficient evidence that by adopting these strategies, some incredible high-performance functional materials have been synthesized artificially. For instance, the synthesis of magnetite by magnetotactic bacteria under ambient conditions has greatly inspired the scientific community. Given the increasing demand for magnetic nanomaterials, a simple and scalable route for the synthesis of size- and shape-controlled magnetite nanocrystals is highly desirable. Scientists have created magnetite nanocrystals with exquisite morphology and uniformity by employing Mms6 protein extracted from magnetotactic bacteria. Cobalt-doped magnetite nanocrystals have also been prepared using this method. Magnetite formation has even been achieved on flat silicon substrates featuring assemblies of Mms6 protein on their surfaces. These methodologies hold great potential for high-density data storage devices [14].

Biomineralization of calcium carbonate by mollusks and sea urchins has also been extensively investigated. As discussed earlier, macromolecules like proteins, glycoproteins, and polysaccharides play a crucial role in the nucleation and crystal growth of calcium carbonate. These macromolecules impart strict molecular control over the selection of specific polymorphs, remarkable morphology, and excellent mechanical properties. Inspired by these molecularly controlled processes, material scientists have generated similar functional materials. For example, thin films of calcium carbonate have been fabricated on the surfaces of organic matrices functionalized with acidic polymers. These acidic molecules induce calcite phase formation with a high degree of morphological control [31]. Calcium carbonate crystals with controlled polymorphs, crystallographic orientations, and various morphologies have also been formed using designed templates and additives with specific functional groups. Unidirectional polymer-calcium carbonate composites have been fabricated by employing macromolecular templates and acidic polymers [14].

Another inspiring feature of biomineralization is the ability of living organisms to produce damage-tolerant structural materials with desired functionality. They achieve this by utilizing amorphous materials as building blocks to produce complex crystalline materials with isotropic mechanical properties. Anisotropic mechanical properties associated with crystalline materials make them vulnerable to fracture. One such amorphous material is amorphous calcium carbonate (ACC), which sea urchins efficiently use to form strong spines [32]. However, ACC is inherently unstable toward crystallization, so developing highly efficient and biodegradable stabilizing agents is highly desirable. Inspired by the ACC stabilized in sea urchin spicules, scientists used cellulose nanofibrils to stabilize ACC. The resulting cellulose nanofibril-ACC composite exhibits high tensile strength and self-healing properties [32]. Additionally, the synthesized composite revealed high humidity sensitivity which could be used for respiration monitoring.

From the depths of oceans to our own mouths, teeth tackle tough challenges in various ways, providing further inspiration for scientists to create biodegradable fibers capable of withstanding stretching. Such fiber materials present opportunities to replace non-biodegradable fibers and components that pollute the environment. Scientists have crystallized the goethite phase of iron oxide, the main component of limpet teeth—one of the hardest biomaterials—where goethite is held in place by chitin polysaccharide. In vitro synthesis of goethite has been carried out in the presence of enzymes secreted by the limpet radula. The resulting material closely resembles limpet teeth. Most importantly, this method works at 14°C and almost neutral pH, unlike conventional methods requiring heating in strong acidic conditions [33].

Another fascinating biomineral is squid ring teeth, which inspire scientists with their self-healing abilities. While capturing prey, squid ring teeth experience fluctuating forces that cause damage. However, unlike human teeth, squid ring teeth demonstrate a remarkable ability to recover from damage. This is attributed to their unique structural features, consisting of a repeating assembly of specific amino acids that form a strong network of hydrogen bonds. These bonds break and reform repeatedly, creating a self-healing effect. Scientists have created self-healing materials by employing squid genes. Being strong, biodegradable, and self-healing, fibers drawn from them could be used for clothing. Astonishingly, these materials also exhibit triboelectric properties, i.e., the ability to create and hold static charges. This property can be exploited to capture particles from the air, serving as high-quality air purifiers [33].

Our teeth also serve as an interesting example, with their unique bilayer structure. In teeth, hydroxyapatite is mineralized within a collagen matrix. Interestingly, the mineralization occurs in layers, with the outer layer containing a higher amount of hydroxyapatite existing as vertical rods. The inner layer is somewhat softer, consisting of horizontal rods with less hydroxyapatite. This unique formation ensures that cracks originating in the outer layer do not propagate deeply but stop at the interface. If teeth were made of pure hydroxyapatite, they would be very brittle. Inspired by this bilayer toughness, heavy-duty steel tools have been manufactured by applying quick heating and cooling treatments to the outer layer, making it harder and better supported by the inner layer [33].

Another way to create tooth-like bilayer structures is through the slip-casting approach. This method begins with the crystallization of magnetic materials under the influence of a weak magnetic field. After forming the outer layer, the process is repeated to create the inner layer, and finally, the two layers are cross-linked through polymerization. An additional advantage of this method is the ability to orient functional materials. For instance, boron nitride composites produced using this approach exhibit anisotropic heat transfer, which could be useful in the electronics industry to cool densely packed components [33].

The approaches discussed above, in conjunction with 3D printing technology, could imitate the role of living cells in forming composite structures. This advanced technology could be used to create artificial organs with significant implications in various fields. Furthermore, machine learning could aid in the endeavor to create new high-performance functional materials. The properties of bio-inspired materials could also be fine-tuned by subtle variations in synthetic conditions. For example, using different polysaccharides or glycoproteins during synthesis could yield new materials with entirely novel properties.

We are surrounded by incredible natural materials with hierarchical structures, exquisite morphologies, and functional properties that surpass our current manufacturing capabilities. However, by drawing inspiration from nature, groundbreaking advancements in diverse areas—ranging from medicine and engineering to energy and environmental sustainability—can be achieved.

REFERENCES

1. S. Mann, Biomineralization: *Principles and Concepts in Bioinorganic Materials Chemistry*, Oxford University Press, Oxford, 2001.
2. H. A. Lowenstam, S. Weiner, *On Biomineralization*, Oxford University Press, New York, 1989.
3. M. Cusack, A. Freer, Biomineralization: elemental and organic influence in carbonate systems, *Chem. Rev.* 2008, 11, 4433–4454.
4. M. Hildebrand, Diatoms, biomineralization processes, and genomics, *Chem. Rev.* 2008, 108, 4855–4874.
5. D. Faivre, D. Schuler, Magnetotactic bacteria and magnetosomes, *Chem. Rev.* 2008, 108, 4875–4898.

6. E. I. Ochiai, Biomineralization: principles and applications in bioinorganic chemistry-V, *J. Chem. Edu.* 1991, 68, 627.
7. S. Mann, Biomineralization: the hard part of bioinorganic chemistry! *J. Chem. Soc. Dalton Trans.* 1993, 1–9.
8. L. Addadi, S. Weiner, Control and design principles in biological mineralization, *Angew. Chem. Int. Ed. Engl.* 1992, 31, 153–169.
9. R. Crichton, Chapter 19 – Biomineralization. In *Biological Inorganic Chemistry* (Third Edition), Elsevier, Amsterdam, The Netherlands, 2019, Pages 517–544, ISBN 9780128117415.
10. F. C. Meldrum, H. Colfen, Controlling mineral morphologies and structures in biological and synthetic systems, *Chem. Rev.* 2008, 108, 4332–4432.
11. R. B. Frankel, D. A. Bazylinski, Biologically induced mineralization by bacteria, *Rev. Mineral Geochem.* 2003, 54, 95–114.
12. F. H. Wilt, Biomineralization of the spicules of sea urchin embryos, *Zool. Sci.* 2002, 19, 253–261.
13. J. Aizenberg, S. Albeck, S. Weiner, L. Addadi, Crystal-protein interactions studied by overgrowth of calcite on biogenic skeletal elements, *J. Crys Growth* 1994, 142, 156–164.
14. A. Arakaki, K. Shimizu, M. Oda, T. Sakamoto, T. Nishimura, T. Kato, Biomineralization-inspired synthesis of functional organic/inorganic hybrid materials: organic molecular control of self-organization of hybrids, *Org. Biomol. Chem.* 2015, 13, 974.
15. D. Faivre, D. Schuler, Magnetotactic bacteria and magnetosomes, *Chem. Rev.* 2008, 108, 4875–4898.
16. M. Tanaka, E. Mazuyama, A. Arakaki, T. Matsunaga, MMS6 protein regulates crystal morphology during nano-sized magnetite biomineralization in vivo, *J. Biol. Chem.* 2011, 286, 6386.
17. D. Kim, B. Lee, S. Thomopoulos, Y. S. Jun, The role of confined collagen geometry in decreasing nucleation energy barriers to intrafibrillar mineralization, *Nat. Commun.* 2018, 9, 962.
18. A. M. Rabie, A. Veis, An immunocytochemical study of the routes of secretion of collagen and phosphophoryn from odontoblasts into dentin, *Connect Tissue Res.* 1995, 31, 197–209.
19. Y. Politi, T. Arad, E. Klein, S. Weiner, L. Addadi, Sea urchin spine calcite forms via a transient amorphous calcium carbonate phase, *Science* 2004, 306, 1161–1164.
20. A. L. Soldati, D. E. Jacob, P. Glatzel, J. C. Swarbick, Element substitution by living organisms: the case of manganese in mollusc shell aragonite, *J. Geck, Sci. Rep.* 2016, 6, 22514.
21. S. Weiner, L. Addadi, Design strategies in mineralized biological materials, *J. Mater. Chem.* 1997, 7, 689–702.
22. G. Falini, S. Albeck, S. Weiner, L. Addadi, Control of aragonite or calcite polymorphism by mollusk shell macromolecules, *Science* 1996, 271, 67–69.
23. N. Tsafnat, J. D. F. Gerald, H. N. Le, Z. H. Stachurski, Micromechanics of sea urchin spines, *PloS One* 2012, 7, e44140.
24. J. Aizenberg, DA Muller, J. L. Grazul, D. R. Hamann, Direct fabrication of large micropatterned single crystals, *Science* 2003, 299, 1205–1208.
25. J. R. Young, S. A. Davis, P. R. Bown, S. Mann, Coccolith ultrastructure and biomineralisation, *J. Struc. Bio.* 1999, 126, 195–215.
26. A. P. Jackson, J. F. V. Vincent, R. M. Turner, Comparison of nacre with other ceramic composites, *J. Mater. Sci.* 1990, 25, 3173–3178.

27. P. Das, S. Schipmann, J. M. Malho, B. Zhu, U. Klemradt, A. Walther, Facile access to large-scale, self-assembled, nacre-inspired, high-performance materials with tunable nanoscale periodicities, *ACS Appl. Mater. Interfaces* 2013, 5, 3738–3747.
28. F. Nudelman, N. A. J. M. Sommerdijk, Biomineralization as an inspiration for materials chemistry, *Angew. Chem. Int. Ed.* 2012, 51, 6582–6596.
29. D. C. Popescu, M. M. J. Smulders, B. P. Pichon, N. Chebotareva, S. Y. Kwak, O. L. J. van Asselen, R. P. Sijbesma, E. DiMasi, N. A. J. M. Sommerdijk, Template adaptability is key in the oriented crystallization of $CaCO_3$, *J. Am. Chem. Soc.* 2007, 129, 14058.
30. Y. Y. Kim, K. Ganesan, P. Yang, A. N. Kulak, S. Borukhin, S. Pechook, L. Ribeiro, R. Krçger, S. J. Eichorn, S. P. Armes, B. Pokroy, F. C. Meldrum, An artificial biomineral formed by incorporation of copolymer micelles in calcite crystals, *Nat. Mater.* 2011, 10, 890.
31. N. Hosoda, T. Kato, Thin-film formation of calcium carbonate crystals: effects of functional groups of matrix polymers, *Chem. Mater.* 2001, 13, 688.
32. W. Wu, Z. Lu, C. Lu, X. Sun, B. Ni, H. Colfen, R. Xiong, Bioinspired stabilization of amorphous calcium carbonate by carboxylated nanocellulose enables mechanically robust, healable, and Sensing biocomposites, *ACS Nano.* 2023, 17, 6664–6674.
33. I. L. Guillou, *A New Generation of Materials Inspired by Teeth*, Chemistry World, Thomas Graham House, Cambridge, 2023.

Index

Note: **Bold** page numbers refer to tables and *italic* page numbers refer to figures.